Global Struggles and Social Change

Global Struggles and Social Change

From Prehistory to World Revolution
in the Twenty-First Century

CHRISTOPHER CHASE-DUNN
AND PAUL ALMEIDA

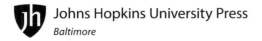

Johns Hopkins University Press
Baltimore

Johns Hopkins University Press
2715 North Charles Street
Baltimore, Maryland 21218-4363
www.press.jhu.edu

Library of Congress Cataloging-in-Publication Data

Names: Chase-Dunn, Christopher K., author. | Almeida, Paul, 1968– author.
Title: Global struggles and social change : from prehistory to world
    revolution in the twenty-first century / Christopher Chase-Dunn and Paul Almeida.
Description: Baltimore : Johns Hopkins University Press, 2020. |
    Includes bibliographical references and index.
Identifiers: LCCN 2019047550 | ISBN 9781421438627 (paperback) |
    ISBN 9781421438634 (ebook)
Subjects: LCSH: Social change—History. | Social movements—History. |
    Right and left (Political science)
Classification: LCC HM831 .C495 2020 | DDC 303.4—dc23
LC record available at https://lccn.loc.gov/2019047550

A catalog record for this book is available from the British Library.

*Special discounts are available for bulk purchases of this book. For more information, please
contact Special Sales at specialsales@press.jhu.edu.*

Johns Hopkins University Press uses environmentally friendly book materials, including recycled
text paper that is composed of at least 30 percent post-consumer waste, whenever possible.

Contents

....

Tables and Figures

Preface

This book was a collaborative endeavor with the two authors contributing equally. Almeida authored much of the introduction and chapters 2, 3, and 6, while Chase-Dunn wrote chapters 1, 4, and 5.

We have a number of people to thank. The staff and editors at Johns Hopkins University Press were patient and delightful while negotiating the arrival of various manuscript drafts over the past four years. They include Brendan Coyne and our original editor, Elizabeth Demers, as well as Lauren Straley. Laura Davulis, Esther Rodriguez, and Kyle Kretzer shepherded the manuscript into final peer review and production. We also benefited enormously from four anonymous external readers who volunteered copious pages of critical commentary that we believe made the final product much more solid. The generosity of the reviewers' time and insights is truly remarkable in an academic world that rarely rewards or acknowledges this hidden but indispensable labor.

Several research assistants contributed with data collection and technical support. They include Maria Mora, Amalia Pérez Martín, Samuel Álvarez, Valezka Murillo, and Jaqueline Novoa. We also appreciate the support of our colleagues and graduate and undergraduate students in the sociology departments at the University of California, Merced, and the University of California, Riverside. At the University of California, Merced, Almeida benefited from the warm environment of faculty in his department, including Sharla Alegria, Camila Alvarez, Irenee Beattie, Kyle Dodson, Charlie Eaton, Edward Flores, Tanya Golash-Boza, Laura Hamilton, Yang Lor, Whitney Pirtle, Zulema Valdez, Nella Van Dyke, Simon Weffer, Elizabeth Whitt, and Marjorie Zatz. He also appreciates encouragement from his graduate students Maria Mora, Rodolfo Rodriguez, Amalia Pérez Martín, Alejandro Zermeño, and Luis Rubén González.

We would like to thank Peter Grimes, E. N. Anderson, Ellen Reese, Marilyn Grell-Brisk, Hiroko Inoue, Rebecca Alvarez, Sando Nagy, and Vladimir Borel for their help in thinking about world revolutions and the ways in which social movements have shaped human sociocultural evolution. We also wish to acknowledge the valuable exchanges we have had with fellow globalization scholars William Robinson, Jackie Smith, Patrick Bond, Valentine Moghadam, Samuel Cohn, Andrew Jorgenson, Sahan Savas Karatasli, Beverly Silver, and Leslie Gates. And this book would not have been possible without the emotional and loving support of our families.

Introduction

In the early decades of the twenty-first century, an international movement to slow the pace of climate change has mushroomed across the globe, reaching small island states in the South Pacific and the most impoverished countries in the interior of sub-Saharan Africa. The self-proclaimed "Climate Justice movement" urges immediate action in reducing carbon emissions and new bold policies to address global warming before irreversible and catastrophic damage threatens the habitability of the planet. By October 2014, during global climate talks at the United Nations, climate justice activists successfully organized 400,000 people in an enormous street demonstration in New York City while also coordinating thousands of collective events around the world. The October 2014 climate campaign was perhaps the most extensive international mobilization in history, and these efforts repeated themselves in November 2015 with the global mobilizations leading to the Paris Climate Accords as well as in September of 2019 with Fridays for Future planetary-wide protest demonstrations. This global movement infrastructure may be humanity's best hope to sustain progressive transnational collective action into the 2020s and push us on an alternative path for ecological survival (Almeida 2019a).

On another front, since the 1980s, multiple waves of resistance have occurred around the globe against the uneven transition from state-led development to neoliberalism—the process of moving from government-directed to unregulated market-driven economies. These waves of opposition have been characterized by mass mobilization against subsidy cuts, price increases, privatization, and trade liberalization. The collective actions are largely driven by the loss of social citizenship rights of access to the welfare state that expanded at an unprecedented rate in the mid-twentieth century but continues to contract in the twenty-first century (Somers 2008). After ethnic and religious-based strife, societal opposition to neoliberal reforms has generated the largest demonstrations and rallies and sustained novel transnational movements such as the World Social Forum (WSF). The social movement struggles against neoliberalism have also eventuated in the rise of post-neoliberal political parties and governments in Latin America and Europe attempting to offer alternatives to free market forms of globalism. Increasingly, free trade globalization is also sparking rightist and xenophobic mobilization throughout Europe, North America, and beyond (Standing 2011; Robinson 2014; Muis and Imerzeel 2017; McVeigh and Estep 2019).

Both Climate Justice and Anti-Austerity movements represent the urgency of understanding how global change affects the ability of citizens around the world to mobilize and protect themselves from planetary warming and the loss of social protections granted in earlier eras. Throughout this work, we address how global change stimulates the formation and shape of such movements. We contend that global economic shifts condition the pattern of social movement mobilizations around the world. These trends are traced back to premodern societies whereby severe disruptions of indigenous communities led to innovative collective actions, often headed by spiritual and charismatic leaders as well as elites. Our approach also examines the influence of global change processes on local, national, and transnational social movements, and how, in turn, these movements shape global institutional shifts. Such an approach can be labeled structural institutionalist with the aim of showing the interplay of global change–social movement interactions over the long run—what Tilly (1989) refers to as "big structures, large processes, huge comparisons."

While most macro-level social science tends to favor methodological nationalism (Beck 2007), we favor a truly global approach. Even as we consider temporal, cross-national and subnational variations in chapters 2 through 5, the focus remains on global-level patterns and processes, from climate change and the deepening of neoliberalism to the spread of authoritarian populism. Local communities and nation-states are embedded in global structures and, at critical historical junctures, mobilized localities and states shape institutions on a worldwide scale. Hence, our approach moves beyond nation-state-centered models of globalization.

Why Globalization and Social Movements Are Critical to Understanding How Social Change Occurs

Dramatic changes observed at the local and national levels of social life often originate in global society (Meyer 2010). Shifts in global economic relations and exchange emanating from powerful actors in the world-system shape stratification and inequality within countries and communities (Chase-Dunn 1998), including ecological disparities (Jorgenson 2012). Fault lines of global, national, and local stratification over environmental rights and economic distribution set the stage for societal conflict and social movement–type mobilizations. When these cleavages reach the point of making people worse off if they fail to act and groups acquire the associational power to resist, we predict heightened levels of social movement activity (Almeida 2018). At times, collective mobilization may succeed and reduce the original level of harm and hardship driving the actions for relatively positive outcomes (Amenta, Andrews, and Caren 2018). In extraordinary circumstances, social movements may alter the structure of power and stratification in general in what we call world revolutions (the emphasis of chapter 4).[1] These are fundamental premises on which we base our analyses throughout this book.

In an age of increasing global integration, understanding the consequences of macro-structural shifts on social and environmental welfare acts as one of the most pressing issues for contemporary social science to address. In this work, we explain the varying capacity of groups and communities to collectively mobilize to defend themselves from past and emerging global threats. Only by analyzing how global economic change and stratification translate into social movement activity and reforms can we contribute to knowledge on how the actions of a mobilized civil society transform existing practices toward a more sustainable planet.

The Language of Social Change

In order to tackle these issues, we employ a variety of sensitizing concepts that play vital roles in the chapters ahead to explain globally induced change. These terms include globalization, the world-system, neoliberalism,

Global North and South, the welfare state, social citizenship, and social movements. *Globalization* is the catchphrase of the twenty-first century used by scholars, government officials, activists, and even popular cultural icons (James and Steger 2014). We define the term in this work as the increasing density and spatial scale of economic and social relations occurring among countries, groups, institutions, and organizations throughout the world (Chase-Dunn, Kawano, and Brewer 2000). We acknowledge that the process of global integration has been uneven over the past several centuries with periods of expansion and contraction (Chase-Dunn 1999). More specifically, we distinguish between (1) structural globalization (spatial integration) and (2) the "globalization project," a neoliberal political and economic policy ideology advocating free international trade, deregulation, and reductions in social welfare expenditures that emerged in the 1970s (McMichael 2016).

By *world-system* we mean the interaction of states through trade, alliances, and conflict in a multidirectional structure stratified by peripheral and core relations (Chase-Dunn et al. 2015). The world-system structure shapes many of the core social movements active across the contemporary world (Smith and Wiest 2012). *Neoliberalism* is a term to describe a largely free market and less-regulated form of capitalism that predominates in the twenty-first century. It is contrasted to an earlier period of state-led development whereby governments regulated large-scale economic activity, planned production, and attempted in various forms to provide a wider social safety net to protect society from economic and environmental harms (Chase-Dunn 2006; Almeida 2016). The world-system origins of neoliberalism are found in two currents. First, the fiscal crisis of the state in the Global North in the 1970s and aggressive lobbying tactics of business networks to deregulate the economy provided one pathway (Dreiling and Darves 2016). The other current derives from the third-world debt crisis of the 1980s that forced large swathes of the Global South to enact austerity and privatization, resulting in neoliberal states by the 1990s (Walton and Seddon 1994).

Scholars of globalization and development now often divide the world into the Global North and Global South. The Global North is characterized as the wealthier geographic region of the world where the most industrialized economies sponsor high-technology research and sophisticated military infrastructures. The Global South refers to lesser developed nations of the world with greater concentrations of poverty and weaker economies. Though far from an exact match, the majority of wealthier countries are found in the northern hemisphere while a greater proportion of the poorer nations are located in the southern hemisphere. There is also tremendous variation within the grouping of Global South countries. The Global South includes the newly industrialized countries of Argentina, Mexico, Brazil, South Africa, Turkey, India, South Korea as well as wealthy petro-states such as Saudi Arabia, Qatar, and Venezuela. But it also includes extremely poor and rural countries such as Haiti, Honduras, and Mozambique.[2] The *welfare state* is a set of institutional practices by governments to provide social protections to the citizens under its rule. This includes public education, access to medical and health care, unemployment benefits, food subsidies for the poorest segments, social security, disability services, and general protection from environmental and occupational hazards (Garland 2016).

The systemic causes of the rise of the neoliberal globalization project are important for understanding waves of capitalist globalization and waves of resistance to them. The saturation of the world market demand for the products of the post–World War II upswing and the constraints on capital accumulation posed by business unionism and the political entitlements of the welfare states in core countries caused a profit squeeze that motivated large firms and investors and their political helpers to try to break out of these constraints (O'Connor 1973). Japanese and German manufacturing, decimated in World War II, eventually caught up with US manufacturing in the 1970s, producing a situation of overcapacity relative to effective demand (Brenner 2002). The possibilities for global investment opened up by new communications and information technology created new maneuverability for capital (Dicken 2015). The demise of the Soviet Union added legitimacy to the revitalized ideology of the free market and this ideology swept the Earth. Not only Reagan and Thatcher, but Euro-communists and labor governments in both the core and the periphery, adopted the ideology of the "lean state"; deregulation, privatization, and the notion that everything must be evaluated in terms of global efficiency and competitiveness (Harvey 2005; Mudge 2018). It should also be noted that the neoliberal globalization project adopted some of the anti-statist ideology of the New Left as well as some of its protest tactics, reminding us of the interactive nature of social movements.

In later chapters, we also use the concepts of *global culture* and *geo-culture*, as they have been developed by globalization and world-system scholars. The modern world-system remains multicultural with different languages, religions, and civilizational meaning systems, but over the past several hundred years these local, national, and regional cultures have interacted with one another and become more similar because of the economic and political interactions and migrations that have occurred. These converging cultures can be understood as composing an emerging world culture in which meaning systems and institutionalized techniques for translation facilitate global hybridity. An example is so-called world music. The process of global culture formation has been heavily influenced by unequal power in the world-system. English has become the most global language because of the successive hegemonies of Britain and the United States. The (nearly) worldwide adoption of English mapping and time-reckoning conventions, with the observatory at Greenwich (down the Thames from London) claiming and successfully obtaining the status of ground zero, is also a reflection of the history of global power. But the Global South has also contributed to the formation of global culture because competition among elites for power has required the ability to communicate with and to understand the peoples of the Global South in order to colonize them and for exploiting their natural resources (Rodney 1981).

The geoculture is the expressly political part of global culture. Our conception of world politics, which is the basis of struggle for increasingly transnational social movements, understands the geoculture that has emerged since the nineteenth century in terms of a tripartite division in which an evolving Centrist Liberalism has been flanked by a coevolving Global Left and a coevolving Global Right. We describe the evolution of these global political formations in chapter 5.

Our focus on the global polity also employs the idea of civil society. In the social science literature, "civil society" usually refers to social organizations that are not part of states and not part of private entities, such as the family, clubs, and other voluntary associations. We use the term this way when we are discussing how civil societies within nation-states are influenced by globalization; but we also discuss "global civil society" in our effort to comprehend global social movements (Kaldor 2003). Contemporary "global civil society" is composed of all the individuals and groups who knowingly orient their political participation toward issues that transcend local and national boundaries and who try to link up with those outside of their own home countries in order to have an impact on local, national, and global issues.[3] The New Global Left is that subgroup of global civil society that is critical of neoliberal and capitalist globalization, corporate capitalism, and the exploitative and undemocratic structures of global governance (Santos 2006; Steger et al. 2013). The larger global civil society also includes defenders of global capitalism and of the existing institutions of global governance as well as other challengers of the current global order.

Valentine Moghadam (2009) reviews the recent literature on "global civil society." She points out that some of the descriptions of the members of global civil society prefer to think of popular movements striving for greater equality as the main players, leaving out those movements that are trying to defend and preserve restrictionist institutions. Moghadam advocates an empirical rather than a normative approach that recognizes the strong participation of very different kinds of movements. She presents an in-depth analysis of radical Islam as a predominantly restrictionist social movement. It is important to recognize that there are very different kinds of players in the global arena of politics, and that this is not a new development. Fascism was an important global movement in the early decades of the twentieth century (see chapter 4). Robert Schaeffer's (2014) typology of global social movements grasps well the complex nature of global civil society. He describes "aspirational" movements in which people are seeking citizenship rights for themselves (Mora et al. 2018), altruistic social movements in which some people seek to assist others in obtaining citizenship rights, and restrictionist social movements that seek to maintain coercive control over some groups of people. Schaeffer also has an admirably broad and useful definition of social movements that include political parties and other organizational actors that are influential in world politics (Almeida 2010a). This is the same comprehensive approach that is employed by those analysts of global politics who have been inspired by the works of Antonio Gramsci (Gill 2000, 2003; Sanbonmatsu 2004).

In our view, the modern world-system is an arena of both political struggle and economic competition over the past several centuries. This implies that global civil society has existed all along. Global civil society includes all the actors *who consciously participate in world politics—those who see the whole of humanity as the focus of their goals*. In the past, it has consisted primarily of statesmen, religious leaders, scientists, financiers, and the owners and top managers of chartered companies such as the Dutch and British East India Companies. This rather small group of elites already saw the global arena of political, economic, military, and ideological struggle as their arena of contestation (Braudel 1984). Elites led religious and secular social movements in which masses were sometimes mobilized, as in the Protestant Reformation. And counter-movements such as the Catholic Restoration emerged in which transnational organizations such as the Society of Jesus (Jesuits) took the form of an early instance of a consciously global political party (Chase-Dunn and Reese 2007). The republic of letters and transnational scientific societies that emerged in eighteenth-century Europe composed a layer of global civil society that Whiteneck (1996) and other scholars have called an "epistemic community." There has been a "Global Left" and transnational social movements led by non-elite actors at least since the world revolution of 1789. Though the Haitian revolution of 1804 was mainly a revolt of slaves on the sugar plantations in Haiti, some of the leaders were literate former slaves that were inspired by the ideas of the French Revolution (James 1963). The Black Jacobins of the Haitian revolution, by depriving Napoleonic France of important sources of food and wealth, played a role in the rise of British hegemony (Santiago-Valles 2005).

While global civil society is still a small minority of the total population of the Earth, the falling costs of communication and transportation have enabled more and more non-elites to become transnational political actors and have increased the extent to which local revolts are able to communicate and coordinate with one another (Almeida and Chase-Dunn 2018). The decreasing costs of long-distance communications and transportation were now allowing some non-elites to play a more important and direct role in world politics. These developments ramped up during the Age of Extremes, the first half of the twentieth century. Internationalism in the labor movement had emerged in the second half of the nineteenth century. Global political parties were becoming active in world politics, especially during and after the world revolution of 1917. The Communist International (Comintern) convened large conferences of representative from all over the globe in Moscow in the early years of the 1920s.

But local revolts in which actors were oriented toward local rather than global power structures have always played a role in world revolutions to the extent that colonial powers reacted to them. Global consciousness is not necessary for global consequences. An objective global interaction network of indirect connections existed long before most people became aware of it. But the spread of global consciousness has made globalization and the local/global imaginary an increasingly important and contentious aspect of world politics.

While social welfare counts a long history from the feudal era to early industrial capitalism (Polanyi 1944), the welfare state took off between the 1930s and 1970s as a global trend. The actual benefits received by citizens from the welfare state we refer to as *social citizenship* (Marshall 1950; Somers 2008). The rights of social citizenship have eroded in tandem with the contraction of the welfare state over the past four decades. In reaction to the threats of the loss of social citizenship rights, groups around the world have formed *social movement* campaigns. Using the recent definition by Snow et al. (2018: 10), social movements are defined as "Collectivities acting with some degree of organization and continuity outside of institutional or organizational channels for the purpose of challenging or defending extant authority, whether it is institutionally or culturally based, in the group, organization, society, culture, or world order of which they are a part." This definition and accompanying concepts are integrated into our larger theoretical perspective of global threats and social movements.

Global Change, Threat, and Structural Equivalence

Our perspective on globalization-induced movements begins with the negative conditions driving collective action. We refer to these conditions as *threats* (Tilly 1978; Goldstone and Tilly 2001). Global forms of neoliberalism and environmental harms act as threats to societies and communities around the world. They also unify groups into campaigns of collective action to reduce the perceived hardships (Almeida 2018). Part of the explanation for the upsurge in social movement activity over the past few decades resides in how these threats are increasingly spread across countries and localities in similar ways. In other words, communities experience similar types of threats of economic austerity and environmental decline. In order for communities to act on particular threats, two other conditions usually need to be present: (1) perception of threat and (2) organizational capacity. Hence, not all threats (even very obvious and dire ones) eventuate in attempts at collective action (Auyero and Swistun 2009). Impacted social groups must come to a shared understanding that a defined threat exists and that they will be made worse off if they fail to act. At times, this comprehension comes together through community interaction. At other times influential elites may bring attention to the threat. As we will see in chapter 5, right-wing movements and populist demagogues often socially construct threats that may have no empirical basis to motivate collective action (Almeida and Van Dyke 2014).

Similarity of threatening economic and environmental conditions across nations and communities is referred to as *structural equivalence* (Strang and Soule 1998). Structural equivalence assists in our understanding of why so many movements in different places are struggling over similar issues such as privatization of the public sector or global warming.

As we will explore in further detail in the subsequent empirical chapters of this book, threats and their perception by civil society are not enough to catalyze collective mobilization. Communities that form social movements also require some level of organization and past strategic experience in mobilizing in order to launch a campaign of social resistance. Demarcating the precise kinds of organizational assets and associational power necessary for collective mobilization will also assist in explaining why some countries and localities resist global threats while others fail to mount a campaign. Hence, threats most likely activate social movement–type mobilization when they are widely perceived in the community or socially constructed by elites and a resource infrastructure is available to sustain collective action against the threat or disruption. We examine the issues of global threats and civil society's organizational capacity at the local, national, and transnational levels of social life over long periods of human history.

In the following chapters, we discuss the scholarly debates and present original research on local, national, and international level variations in social movement mobilizations in relation to global change. At the local and subnational levels, scholars are beginning to recognize that, while there are homogenizing impacts of globalization at the macrolevel such as global scripts adopted by individual nation-states, global intrusions within nations and localities are highly uneven (Sassen 2008). The "glocalization" literature sheds light on how global-change processes are interpreted at the local community level and the likelihood for collective mobilization (Auyero 2002; Almeida 2012). The vast literature on recent anti-neoliberal mobilizations around the world over the past two decades provides the most comprehensive body of research to summarize the key conditions associated with national level and local level social movement activity in response to the global transition from state-led development to neoliberalism (the theme of chapter 2).

Themes and Structure of Global Change and Social Movements

This book employs an innovative perspective on social movements in response to globalization processes. Four areas of global change and movements are examined: (1) historical and theoretical perspectives on globalization and social movements, including premodern movements; (2) research on local and national challenges to globalization; (3) scholarship on transnational social movements; and (4) debates on the consequences and outcomes of collective mobilization in response to globalization. Special attention is also given to three important movements that have been induced by globalization in the Global North and South: anti-neoliberal mobilizations, climate justice, and right-wing populism.

We begin by providing historical and theoretical context to global change and social movements. Chapter 1 analyzes the early history of global change and collective movements. It offers a compelling framework demonstrating that waves of social movements and world revolutions can be traced back to premodern polities. Such an approach provides an important corrective to extreme globalization perspectives that give the impression that global counter-movements originate in the late twentieth century. Studies of earlier periods in which clusters of local movements broke out across world regions—so-called world revolutions—also shed light on the contemporary waves of global protest (Silver 2003; Martin et al. 2008; Beck 2011; Mason 2013; Schaefer 2014). These works examine the similarities and differences between the clusters of local and transnational social movements that occurred around symbolic years in world history, for instance, 1789, 1848, 1917, 1968, 1989, and 2011 (Almeida and Chase-Dunn 2018).

Our first chapter reaches further back in history to examine collective action in premodern polities. We give special emphasis to how world-system disruptions and threats led to alternative forms of collective action, such as the Ghost Dance. We highlight that social movements are not solely a creation of the nineteenth century with the expansion of states, national parliaments, and capital, but rather collective action dynamics go back to much earlier epochs of early human settlements and civilizations. Among the important features and patterns found in premodern movements are elite sponsorship, heavy emphasis on spirituality and religion, societies undergoing external disruptions and threats, and critical roles played by charisma and emotions. Indeed, as John Markoff (2015a: 69) has summarized in his review of historical analysis and social movement research, "studying movements in the past is a way of enlarging the variety of movements taken into account in our theoretical understandings."

Chapter 2 highlights contemporary processes of globalization and resistance, including the indispensability of the organizational infrastructure established under state-led development. We emphasize our original conceptual scheme of globally induced threats driving collective action at the local and national levels of political life. We place our framework in a larger structural perspective to explain the emergence of the most important global movements of the twenty-first century covered in subsequent chapters. Chapter 2 focuses on globalization and social movements in the Global South with special attention given to movements battling over various forms of neoliberalism, including austerity, price hikes, privatization, and free trade. The organizational capacity of civil society to resist will be shown to be rooted in earlier periods of state-led development. These processes are brought down to the local level with systematic empirical data on recent anti-neoliberal uprisings in Central America.

The chapter also recognizes the resistance to neoliberalism in the Global North with an examination of Anti-Austerity movements, and other related forms of rebellion against labor flexibility and income inequality. We engage the debates regarding the effects of the globalization project on class structures in the Global North and South and the efforts that are being made to solve the massive collective action problems of the global precariat (Standing 2011).

Chapter 3 turns our attention to environmental crisis and globalization. We show the rise of the Climate Justice movement. The movement for climate justice has benefited from the recent rise of transnational collective action campaigns that can coordinate protests against mounting ecological threats of global warming. The Global Economic Justice movement and the Anti-War movement are particularly important organizational predecessors to the Climate Justice movement. Both precursor movements appear as central transnational activist networks in the "neoliberal timescape" at the beginning of the new millennium (Gillan 2018). The work of the global economic justice and anti-war campaigns in the early 2000s set the global activist infrastructure in place for the transnational movement battling global warming to launch in the late 2000s and 2010s. We also present novel global data to demonstrate these relationships. We contend that the Climate Justice movement now stands as the most extensive transnational movement across the globe, with the greatest potential to form and maintain progressive coalitions in the present and coming decades.

While the internationalization of collective struggle is a theme throughout our work, chapter 4 brings the future of transnational movements front and center—as well as the possibility of radical social change in the 2020s and the types of coalitions and social sectors guiding such a struggle. One of the most interesting outcomes of global integration is the double-edged-sword nature of the processes by which new communications technologies have facilitated a greater density of social movement ties among activists and issue advocacy organizations across the globe (Almeida and Lichbach 2003). The past three decades have witnessed a near tripling in the number of transnational social movement organizations (Smith and Wiest 2012)—social movements operating in at least two countries. We examine major transnational social movements of the twenty-first century, including the Global Justice movement and the World Social Forum (WSF). The first author conducted surveys of attendees at four Social Forum meetings (Porto Alegre, Brazil; Nairobi, Kenya; Atlanta, Georgia; and Detroit, Michigan). Analyses of these responses allowed us to understand the demographical and political characteristics of this slice of global civil society and we also use these results to study the relationships among the different movements that are active in the social forum process. We focus on how these movements have sought to overcome the collective action problems of coordinating movement activities across multiple national territories and languages. The WSF is arguably the largest global movement on the left to emerge in the current century, while climate justice is the most extensive in terms of territorial reach and future prospects for growth. We conclude by hypothesizing the potential for different social sectors to enter into progressive coalitions and advocate for systemic global change in the 2020s and beyond.

Chapter 5 examines the roots of contemporary right-wing populism and neofascism in the context of world historical change over the past two centuries. The chapter compares twentieth-century fascism with the neofascist and authoritarian populisms that have emerged in recent decades. Focused attention is given to the ideological battle between liberalism, the Global Right, and the Global Left in the context of waves of globalization and deglobalization.

In our conclusion, chapter 6, we document the consequences of globally induced movements. The lasting impacts of social movements catalyzed by global change are arguably the most crucial dimension to empirically verify. In an era of escalating global risks (Centeno et al. 2015), social movements provide one potential pathway to support global level reforms. In this final chapter, we discuss the status of key contemporary global movements and where future scholarship is most likely to be fruitful. Such cases include alternatives to neoliberalism in the Global South brought about by anti-neoliberal movements, and the influence of global environmentalism on coordinating and sustaining a progressive transnational and rainbow coalitions.

Special emphasis is placed on the ecological threat of global warming and the planetary mobilizing infrastructure developed by the Climate Justice movement. We place this consideration in world historical context by comparing contemporary transnational movements with those that occurred in earlier conjunctures in which the anti-colonial, labor, and feminist movements impacted the global normative and political orders. Readers are also reminded of the persistent role played by religion, charisma, emotions, elites, and threats in generating the dynamic interplay of global change and social movements.

ONE

Social Movements and Collective Behavior in History and Prehistory

....

Collective behavior and social movements have been important drivers of social change since the emergence of human language. In social movement studies in sociology, the presentism of most of the field is sometimes justified by specious distinctions between "modern" and "premodern" movements, thus facilitating ignorance of the roles that social movements and collective behavior have played in human history outside of modern Europe and in prehistory. We contend that social movements and collective behavior have been central to processes of social change since the Stone Age. Theories that claim to explain the long emergence of social complexity and hierarchy must try to comprehend the ways in which deviance, rebellions, revolutions, and other kinds of spontaneous and less institutionalized behavior have played.

Spontaneous group action motivated by strong emotions, typified by the behavior of crowds, has had important consequences for both maintaining and changing social structures since all humans lived in small, nomadic, foraging bands. Ethnographically known hunter-gatherer polities have institutionally reinforced egalitarian structures in which consensual norms organized primarily around kinship promote sharing and reciprocity and punish aggrandizing behavior (Flannery and Marcus 2012; Bowles and Gintis 2013; Chase-Dunn and Lerro 2014). This institutionalized egalitarianism emerged over a long period despite the hierarchical structures present in the bands of the great apes from which humans descended. So how did the relative egalitarianism of early human groups emerge? Explanations for the emergence of equality in human hunter-gatherer bands focus on the ability of group coalitions to control the behavior of dominant individuals (Boehm 1999). Peter Turchin (2016a: chap. 5) contends that the ability of humans to throw projectiles allowed coalitions of smaller people to exercise coercive power over larger and stronger ones. Coercive power of this sort was undoubtedly exercised by mobilizing emotional solidarity (Jasper 2018) among the members of the coalition that was threatening the aggrandizer. This basic structure of coercive power eventually produced a moral order understood as kinship in which institutionalized rights and obligations came to be the main promoter of egalitarian sharing and reciprocity.

Migrations to new locations were usually motivated by population density and competition for resources, but departure events and many other group decisions were legitimated in terms of ideological formulations and disagreements. The emergence of larger-scale leadership, hierarchical kinship, and class formation were often legitimated in terms of discourses about authority, connections with ancestors and disputes about origin myths or the correct performance of rituals. We describe evidence of collective behavior and social movements among hunter-gatherers and discuss premodern movements that are known from historical accounts. And we delineate the similarities and differences between social movements in small-scale systems and in larger, state-based systems.

Social movements have long been studied as instances of what sociologists have called collective behavior (Blumer 1951; Smelser 1962). Collective behavior is understood to consist of noninstitutionalized and somewhat spontaneous actions by human individuals and groups, especially those that occur in crowds, riots, revolts, revolutions, fads, and fashions (Buechler 2011). Most sociologists who study social movements think of them as a modern phenomenon, but many historians have noted the important role played by rebellions in premodern dynastic transitions (Brunt 1971; Richardson 2010; Goldstone 2014; Collins and Manning 2016) and some historical ethnographers (e.g., Spier 1935; Wallace 1956; Lawrence 1964) have noted that premodern small-scale polities also reveal instances of collective behavior that seem rather similar in many ways to the processes and patterns exhibited by modern social movements.[1] Neil Smelser's (1962) general theoretical approach contends that collective behavior and social movements emerge when existing social institutions are doing a poor job of meeting people's needs and expectations and that social movements are important agents of social change. Manuel Castells (2013) offers a similar perspective in the Information Age. This approach is easily extended to premodern polities and suggests that collective behavior and social movements have been important causes of the evolution of social institutions since the Stone Age.[2]

Most sociological theories see religions as mainstays of social structure and stability. Institutionalized moral orders, assumptions about what exists and what is right, are pillars that produce stable expectations and reinforce other institutions. Emile Durkheim (1915) described religions as projections of social structure on the sky (see also Swanson 1960). Egalitarian polities projected beings who had powers, but also quirks and faults, while many hierarchical polities project and worship a single omniscient and omnipotent god of the universe.

But religious ideas have been recurrent matters of contestation in prehistoric, historical, and contemporary politics (Denemark 2008; Moghadam 2013). Religions, even animistic belief systems, are discourses about authority that are often in dispute. Disagreements about creation myths or the correct performances of rituals have often been the ideological forms used to mobilize collective action in both modern and premodern human polities. Projections on the sky may be used to defend older social structures or to propose and justify new ones. And indeed, despite the emergence of secular humanism, religious identities and doctrines continue to be the basis of much hegemonic and counter-hegemonic contestation and mobilization in the contemporary world (Flores 2018). Some social movement theorists have tended to ignore "primitive" social movements that are motivated by religious ideologies or that do not utilize modern repertoires of contention in order to focus only on "modern" secular movements. These latter are more likely to employ frames based on secular humanism and legitimation of authority from below (popular sovereignty). But this approach obliterates prehension of the role that social movements have played in sociocultural evolution and occludes the analysis of those contemporary social movements that still employ religious ideologies and not so civil modes of contention. It also ignores the fact that secular humanism emerged as part of Confucianism long before it made its appearance in the West.

The main idea used to justify not studying premodern social movements claims that premodern or primitive movements were "reactive, backward, and parochial," whereas modern movements are proactive. This is usually conflated with distinctions between those movements who are "functional" because they profess a version of secular humanist ideology from those that employ ideas based on religion. Proactive movements are deemed to be rational efforts to engage and modify political institutions, especially states, whereas reactive movements are more expressive and concerned with ideologically inspired identities. Ho-Fung Hung's (2011: Introduction) careful and insightful study of protests during the mid-Qing dynasty in China provides a useful overview of the teleological distinction between modern and premodern social movements from Marx and Weber and including the work of many of the leading lights in the sociological social movements literature. Charles Tilly's studies of the structural roots of contentious behavior in the West are indispensable, but they downplay the non-West and premodern social movements based on the distinction between reactive versus proactive movements (Tilly and Wood 2013).[3] The problem here is that some Western and recent social movements are reactive in the sense that they do not propose institutional solutions to problems, and non-Western and premodern movements were often engaged with states and sometimes did propose changes in state institutions. So, the distinction as applied to broad periods and world regions, is specious. It is a poor excuse for ignoring non-Western and premodern world history and prehistory. The downplaying of threat-induced or defensive mobilization in the political process literature also tends to highlight proactive movements (Almeida 2018).

Globalization remains a contested concept in both popular discourse and in social science. As mentioned in the introduction, we propose to distinguish between first, globalization as the spatial expansion and deepening of human interaction networks, and second, the "globalization project" that has emerged since the 1970s as a hegemonic discourse about deregulation, privatization, and austerity (Chase-Dunn 2006). Globalization understood as the expansion and intensification of spatial interaction networks has been going on since modern humans migrated out of Africa to occupy the other continents (Massey 2005; Modelski, Devezas and Thompson 2008; Jennings 2010; Chase-Dunn and Lerro 2014; Harari 2015). In this chapter, we discuss the roles that social movements have played in social change in hunter-gatherer bands, world-systems of sedentary foragers, and in the processes of class formation and state formation that occurred in several different world regions. We propose that the social movements have been an important aspect of the processes that have driven the expansion and intensification of human interaction networks since the Stone Age.

Recall from the introduction that we prefer David Snow and Sarah Soule's (2010: 6–7) flexible definition of social movements as "collectivities acting with some degree of organization and continuity, partly outside institutional or organizational channels, for the purpose of challenging extant systems of authority, or resisting change in such systems, in the organization, polity, culture, or world system in which they are embedded."[4] Robert Schaffer (2014) points out that single individuals often start social movements. And John W. Meyer has said that all human behavior is collective because even hermits have society in their heads. This definition is sufficiently flexible to allow for its application to premodern settings in which authority structures were less hierarchical and organizational forms were less bureaucratic. All human polities have had some kind of authority and some kind of organization. Though authority in small-scale polities is less centralized and has less power, contentious disagreements and disputes still occurred and people mobilized themselves and one another to address these conflicts.[5]

Emigration by groups (hiving off) was an important outcome of disagreements as well as a response to population pressure. Decisions to leave were often framed in terms of disagreements about the moral order or correct ritualized behavior. Formal authority structures that are centralized provide a clear and convenient target for protests and so social movements in hierarchical polities are more easily focused on the symbols and structures of power once these have become institutionally defined. As formal organizational forms emerged, social movements were able to utilize these as both targets of protest and as instruments for coordinating participants, increasing the scale and effectiveness of the movements (Tarrow 2011). And new and better technologies of communication and transportation were also used by movements. These changes in scale and effectiveness explain why most social movement theorists define movements as a modern phenomenon, but the less organized, more spontaneous forms of collective behavior that were frequent events in premodern polities also had important effects on changes in legitimating ideologies, regime change, and the development of new techniques of power.

Social movement theorists have also usually assumed that movements come from below and target those above. Consideration of how movements might have operated in egalitarian polities problematizes this assumption; but we contend that it is also a problem in more stratified contexts. Social movement research has long confirmed that leaders of the masses are more likely to have had origins and resources that allowed them the opportunities to mobilize. The poorest and most downtrodden people rarely have the resources that are needed. So, leaders of movements tend to be at least from the middle classes. It is also important to note that elites often sponsor and lead social movements (Auyero 2007). Elites are undoubtedly more likely to utilize institutionalized forms of power because they have greater access to these, and this is one reason why social movement theorists tend to assume that movements come from below. But the activities of elite factions that were seeking to mobilize public opinion and to influence the course of history have often taken forms that are very similar to social movements from below. Elites sometimes seek only to mobilize other elites, but they also often mobilize the masses in their struggles with other elite factions. Either way, these efforts often take on aspects of collective behavior and they should be included in the study of how movements have caused social change or sought to protect the privileges and property of those in power.[6]

It is well-known in the social movements literature that emerging social movements generate counter-movements (Zald and Useem 1987; Meyer and Staggenborg 1996). It is possible that these cancel each other out, but if this always happened nothing would change. Indeed, social change is characterized by many cycles, but there have also been upsweeps in the levels of complexity, hierarchy, and size. Dynastic cycles do not just return to the old normal. Some of the recoveries go on to establish a new normal. The elite factions in the secular cycle model (see below) sometimes led popular coalitions that allowed polities and settlements to become larger than they had previously been. These upsweep coalitions (the alliance during territorial expansion) came out of periods of great contention in which people had become tired of conflict and disorder and were willing to support a new regime that could maintain order and promote recovery. The insight that movements produce counter-movements is not wrong, but it needs to consider the conditions under which this leads to upsweeps instead of simply returning to the former normal.

Collective Behavior in Small-Scale Polities

The sources of evidence about collective behavior processes and premodern social movements in small-scale polities are mostly indirect.[7] We must rely on archaeological evidence regarding those small-scale polities that evolved during the period before the emergence of writing, and because small-scale polities did not themselves have writing, documentary evidence on those that are known historically relies on the reports of people who did have writing: travelers, missionaries, settlers, and eventually professional ethnographers. Knowledge about social movements based on archaeological evidence requires suppositions about the meaning of those behaviors that are indicated by surviving physical evidence. Documentary and ethnographic reports come from observations that are distant in time from the peoples who originally carried out sociocultural evolution in small-scale polities. Some of these people survived until recent centuries and documentary evidence requires the assumption that they are representative of peoples who lived long ago. This is obviously problematic as will be seen in our discussion of social movements among sedentary foragers below.

We know that humans began burying their dead about one hundred thousand years ago. Beads appeared in Southern Africa about seventy thousand years ago (Klein and Edgar 2002) and dramatic cave paintings in Europe are about fifty thousand years old. Despite that humans only arrived in the Americas about fifteen thousand years ago, there is conclusive evidence that nomadic hunter-gatherers were gathering together to build monumental mounds at Watson Brake in Louisiana as early as 3500 BCE (Saunders et al. 2005) with elaborate and rather large-scale mound complexes appearing between 1650 and 700 BCE at Poverty Point, Lousiana (Sassaman 2005).

While sociocultural evolution was slow compared with the accelerating pace of change that we know from recent centuries, big changes did occur. Population density increased as the land filled up with humans. Annual migration routes of nomads became smaller and regional groups developed culturally distinctive tool kits (Nassaney and Sassaman 1995). Sedentism emerged. Trade networks and regional specializations developed. Settlements and polities got larger. Monumental architecture emerged.

Religious ideologies spread across wide regions. Normative regulation was the main form of sociocultural regulation, but this did not produce stasis (Arnold 2004).

Both nomadic and sedentary hunter-gatherers (foragers) are known ethnographically and the cave paintings of some of these are associated with ritual activities.[8] The Chumash were sedentary foragers who lived along the Southern California coast in what is now Santa Barbara and Ventura counties. They built and used a distinctive plank canoe (*tomol*) that allowed them to fish offshore and to develop a trade network that linked those living on the Northern Channel Islands with the villages on the mainland (Arnold 2004). The large coastal villages were also connected by trade in food items with smaller inland villages in the mountains and valleys adjacent to the coast (Gamble 2008). The dramatic cave paintings in the territory of the ethnographically known Chumash are thought to have been associated with the 'antap cult, a secret organization of elites who taught esoteric astronomical knowledge and utilized mind-altering *toloache* (jimsonweed) to gain access to the spirit world (Johnson nd; Romani 1981). Chumash polities were in transition from the more typical egalitarian social structure of other California village-living hunter-gatherers toward class formation. The 'antap cult was a manifestation of this transition in which the emergent elites from autonomous polities demonstrated their superiority over non-elites by carrying out exclusive ritual performances behind outdoor partitions or in isolated remote locations such as Painted Cave, so that commoners could not see them (Hudson 1982).

Ethnographers studying Northern California indigenous polities also observed cults (White Deer among the Hupa, Hesi, Kuksu and Waisaltu among the Patwin, as well as the Bole-Maru and Bole-Hesi) in which songs and dances were spread from group to group by moral entrepreneurs and messengers (Halpern 1988). The most closely studied instance of this phenomenon is the 1870 Ghost Dance. The 1870 Ghost Dance was an earlier version of the more famous 1890 Ghost Dance studied by James Mooney 1965 [1896] and many others (e.g., Wallace 1956; Thornton 1981; Smoak 2006). The 1890 Ghost Dance supposedly originated when god spoke to Wovoka (Jack Wilson) while he was alone in the woods near the Walker River in Western Nevada cutting logs. Wovoka became the Paiute messiah who spread the word from Nevada to many other tribes, provoking a social movement of resistance to settler colonialism. Wovoka's word spread east, exciting the Lakota (Sioux) and other tribes to don ghost shirts that were believed to repel bullets. The ensuing rebellion led to the murder of Crazy Horse and the tragic massacre at Wounded Knee—an ethnocidal slaughter by the army of the white settler colonial state (Mooney 1965; Fenelon 1998: chap. 5; Dunbar-Ortiz 2014). But it was Wovoka's uncle who had spread a very similar doctrine twenty years earlier in the 1870 Ghost Dance.

The earlier 1870 Ghost Dance was studied by Cora DuBois ([1939] 2007). She interviewed surviving participants and observers of this movement as it spread from the Nevada Paiutes near Walker Lake across Northern and Central California and Southern Oregon (see also Spier 1927; Gayton 1930). The most prevalent formulation of the Ghost Dance doctrine was that all the Native Americans who had died in the past were going to return to Earth and the whites would disappear. Du-Bois was told that the Ghost Dance doctrine predicted that all the deceased Native Americans since the beginning of time were returning, coming up from the south, and they would wreak a grim vengeance on the colonizers.[9] Mixed ancestry (metis) would turn into rocks. Nonbelievers would turn into rotten logs. And the future would be a happy world of abundance and restored nature in which sickness and death did not occur.[10]

Moral entrepreneurs carried the "hurry up word" that the Native Americans should come together with certain costumes and songs and dances in order to facilitate the return of their dead loved ones. DuBois tracked the spread and morphological changes of the 1870 Ghost Dance and analyzed why some groups embraced the new ideas while others rejected them. The messengers were "dream doctors" who used horses and wagons to travel to the homelands of other tribes in order to teach the dances and songs and spread the word. Most of the Native Americans did not use written communication and so the dream doctors were reliant on oral communication to spread the ideas of the Ghost Dance.[11] The acceptors were those groups that had been most disrupted by the arrival of Euro-American miners and settlers. The rejecters had been less disrupted and the authority of the traditional shamans was more intact.[12] In the regions that were more remote, and so less disrupted, the traditional shamans were able to convince their co-villagers to reject the dream doctors and the Ghost Dance. In more susceptible locations, local entrepreneurs mixed ideas from the Paiute Ghost Dance with older traditional belief complexes such as the Kuksu cult to produce new hybrids such as the Bole Maru (Big Head) Dance. And so, the form of the rituals and the ideas became modified as they traveled.

The Ghost Dance ideology involved elements that were quite distinct from earlier ritual practices and ideas, and some of these were probably due to the influence of nonindigenous (European) ideologies and practices. Women and children were usually not allowed to participate in the older "dangerous" invocations of spiritual power, whereas they were allowed and encouraged to attend the Ghost Dances. The previous indigenous beliefs of most Native Californians included a fear of the spirits of the dead and careful efforts to ensure their departure to a distant and separate realm so that they would not disrupt the living. The idea that the dead would come back was blasphemy to many of the traditional shamans. Some of the dream doctors who took the word on the road were also economic entrepreneurs who instructed dancers to bring their valuables to the ceremonies to contribute to the cause, and who sold ritual paraphernalia such as chicken-feather capes and charged admission to the dance performances. In some cases, the ideas of the Ghost Dance were remixed with the older cult ideas, producing local variations. And the Ghost Dance ideas were appropriated and further transmitted by local enthusiasts.[13]

One problem with the study of movements that are ethnographically known is that they usually occurred in a context in which indigenous life was being radically altered and threatened by the processes of colonial domination and integration into the expanding modern world-system and so it is difficult to know which elements of the movements were characteristics of precontact polities and which were borrowed from the invading culture. The millenarian aspect of the Ghost Dance—which was a "hurry up word" in which the old world was supposed to be coming to an end and a new world was coming in to being—is often thought to have been such a borrowing from eschatological Christianity. Norman Cohn (1970, 1993) contends that Millenarianism did not exist before its emergence with Zoroastrianism and then it spread to Judaism and Christianity. But it is entirely possible that many human cultures contained both the tropes of a stable cosmos versus an immanent radical transformation. The five suns of Mesoamerican religion, in which the present world was preceded by four other cycles of creation and destruction (Miller and Taube 1993), and related ideas known among village-living hunter-gatherers of indigenous California (Bean 1974) suggest that the notion of a radical cosmological transition was not uniquely invented by Zoroaster. Millenarianism is a powerful trope for mobilizing collective action that continues to play an important role in twenty-first-century social movements (Lindholm and Zuquete 2010). It may be far older and more widespread than Cohn believed. Many human cultures probably contain both notions of eternal order and of imminent transformation.[14]

Anthony F. C. Wallace's (1956) depiction of the 1890 Ghost Dance and cargo cults as "revitalization movements" focused on the ways in which these nativist movements were adaptive responses to the colonial disruption of indigenous cultures that mixed older institutional forms with new elements inspired by contact with the expanding Europe-centered world-system.[15] Cora DuBois ([1939] 2007: 116) has a version of this idea with an interesting twist. Discussing the effects of Christian ideas on the ideology of the dream doctors she says, "it mirrors the accumulating despair of the Indians and their realization that there was no room for them in the new social order. Christian beliefs, which were an outgrowth of a not dissimilar cultural situation, offered a ready-made escape into supernaturalism from realities that had become intolerable because they offered nothing but defeat."

In other words, there was a functional congruence between the ideology of salvation that emerged from an Iron Age Roman colony in Palestine and the situation of indigenous peoples of the Americas.[16] Wallace also surmises that James Mooney's sympathy with the Native Americans was partly due to his support for Irish national independence from British colonialism (Wallace 1965: vi). The 1870 and 1890 Ghost Dances were desperate responses of resistance to settler colonialism that ended in ethnocidal state repression.

While the 1870 and 1890 Ghost Dances were obviously responses to the colonial context, Leslie Spier (1935) studied an earlier Prophet Dance that had emerged among the plateau and coastal groups in what became the Canadian province of British Columbia and the US state of Washington.[17] Spier demonstrated many similarities between the Prophet Dance complex that emerged in the early decades of the nineteenth century with the Ghost Dance doctrines and practices that emerged in the 1870s and the 1890s. He contended that the Prophet Dance complex was not caused by the disruption of indigenous societies by the arrival of the Europeans because not much disruption had occurred at the time of the emergence of the Prophet Dance. Wayne Suttles (1987) agrees and proposes that the Prophet Dance—in which dream doctors proselytized across a wide area based on dances and songs they had learned by visiting the land of the dead—was a response to the need of indigenous polities for larger scale political leadership to mediate expanding indigenous trade networks and competition among polities for control of resources.[18] This was a case in which an autochthonous millenarian social movement constituted an effort to promote the emergence of new and larger forms of authority. We are not contending that all social movements are endogenous causes of evolutionary change. As with the case of the Ghost Dance, external pressures of settler colonialism drive the emergence of defensive collective action. But we do contend that the endogenous evolution that was occurring within world-systems composed of small-scale polities was likely caused by acts of collective behavior that legitimated new ways of life.

Chiefdoms experienced a rise and fall pattern that was somewhat similar in form to that of larger states and empires (Anderson 1994). Some of the rises were the result of conquest by semiperipheral marcher chiefdoms, but others may have been the outcome of a demographic process somewhat similar to the "secular cycles" described for state-based systems by Jack Goldstone (1991) as well as Peter Turchin and Sergey Nefedov (2009). Turchin and Nefedov formalized Jack Goldstone's (1991) model of the secular cycle, an approximately two-hundred-year-long demographic cycle, in which population grows and then decreases. Population pressures emerge because the number of mouths to be fed and the size of the group of elites gets too large for the resource base, causing conflicts and the disruption of the polity. Turchin and Nefedov tested their model on several agrarian empires, confirming the principle that population growth and elite overproduction led to sociopolitical instability within states. We think that somewhat similar processes may have been operating within chiefdom polities.

The main differences between social movements in small-scale polities and those that occur in larger and more complex polities is the social structural context itself.[19] As mentioned above, small-scale human polities were fiercely egalitarian, and integration was mainly based on consensually held norms and values. There were no markets and states to stand behind the moral order. They suppressed aggrandizing behavior by dominant individuals, minimized the inheritance of wealth by burying or burning personal possessions at death, and redistributed use rights to natural resources based on need (Flannery and Marcus 2012; Turchin 2016a: chap. 5). Decisions were made by consensus in discussions that included all adults.

The social movement literature tends to assume the existence of hierarchy and that social movements challenge hierarchical authorities. When there is little or no hierarchy, how is collective action mobilized? The example of the 1870 Ghost Dance provides the answer. Charismatic bearers of new songs and dances and visions of transformation (dream doctors) challenged the ideas and ritual practices of older authorities (traditional shamans) and mobilized action around the new ideas. Even in small-scale egalitarian polities, there were some authorities and there were taken-for-granted practices and ideas that could be challenged. In the case of the revitalization movements, these were adaptive changes to the larger colonial situation, but it is likely that earlier evolutionary reorganizations were similarly implemented by religious social movements as indicated by the example of the Prophet Dance discussed above.

Theocratic Early State Formation

The emergence of socially constructed hierarchy had to overcome great resistance (Scott 2017). There is considerable evidence that hierarchies emerged during periods of warfare and internal strife in which people were motivated to accept the claims of superiority of a chiefly class or a charismatic leader who was able to promise (and deliver) better security. Charisma is a critical feature of collective action past and present. Fatigue from insecurity plus circumscription (the lack of feasible emigration destinations) (Carneiro 1970, 1978) lowered the resistance to claims of superiority on the part of emergent elites. These new elites usually legitimized their actions in terms of reformulated religious ideologies, and they used these to mobilize collective labor toward both monumental and productive projects. As interaction networks and polities got larger and more complex, collective behavior and social movements took forms that were more recognizable by students of modern collective behavior.

Most studies of the emergence of early states depict a situation in which religion played an important role in the establishment of a layer of authority over the top of existing kinship structures. Stephen Lekson (1999) contends that the rise of a relatively large polity with monumental architecture at Chaco Canyon (in what is now Northwestern New Mexico) was organized by a religious elite using astronomical ideas (indicated by the ritual roads centered on the biggest edifice (Pueblo Bonito) in the canyon).[20] The settlement at Chaco Canyon was largely abandoned in the twelfth century and most archaeologists ascribe this to climate change (reduced rainfall and catastrophic floods that lowered the water table by washing out a deep arroyo (e.g., Fagan 1991). Lekson contends that the Chaco priests led a large group to establish another central place directly north of Chaco (Aztec Ruin near the Animas River) and then, after another flooding episode that destroyed an irrigation system, led a third migration to establish Paquimé (Casas Grandes) directly south of both Chaco Canyon and the Aztec Ruin settlement. If Lekson is correct, Chaco elites were employing an astronomical ideology to mobilize their populations to adapt to ecological and climatic conditions by relocating and developing more resilient forms of irrigation.

Timothy Pauketat (2009) presents a thorough overview of the archaeological evidence that has accrued regarding the rise of Cahokia and the Mississippian culture complex of which it was an important center (see also Mann 2005). Pauketat contends that the construction of this large settlement in the American Bottom (now East St. Louis) was organized by religious entrepreneurs who created a dramaturgical set of monuments organized around a religious ideology that attracted large numbers of immigrants and legitimated the mobilization of a huge investment of human labor for the building of the dramatic monumental mounds. Some archaeologists prefer to consider Cahokia to have been a complex chiefdom rather than an early state, but the archaeological evidence of large-scale human sacrifice favors the idea that this was indeed a state.[21] Mound 72 at Cahokia contained the remains of fifty-three decapitated young women, apparently appropriated from poorer families in outlying villages, who seem to have been honored to join a recently deceased king in the afterworld (Pauketat 2009: 133–134; Emerson and Hedman 2016).

In the Late Chalcolithic period (4000–3100 BCE), the first true city (Uruk) grew up on the floodplain of lower Mesopotamia, and other cities of similar large size soon emerged in adjacent locations. This was the original birth of "civilization" understood as the combination of irrigated agriculture, writing, cities, and states. The main architectural feature of these new Sumerian cities was the temple and this structure has long been considered the primary institution of a theocratically organized political economy. Later evidence about Sumerian civilization shows that each city was represented by a god in the Sumerian pantheon and the priests and populace were defined as the slaves of the city god—this justifying the accumulation of surplus product and the mobilization of human labor for building monumental architecture (Postgate 1992). The Sumerian cities erected their states—specialized institutions of regional control—over the tops of kin-based normative institutions (Zagarell 1986). The emergence of these large cities was made possible by the development of irrigation canals that watered large fields for the planting of grain.[22] Assemblies of lineage heads long continued to play an important role in the politics of Mesopotamia (Van de Mieroop 1999).

One interesting difference between the emergence of archaic states in Mesopotamia from other instances of pristine state formation is the apparent absence of ritual human sacrifice. A powerful way to dramatize the power of a king was to bury a lot of other people with him when he died. Except for the Third Dynasty of Ur (the royal cemetery at Ur), there is little evidence of ritual human sacrifice in Mesopotamia.[23] The temple economy required contributions of goods and labor time, including animal sacrifices that were consumed in religious feasts. The sacrifice of humans in Mesopotamia, as with modern states, was mainly confined to killing in battle.

So, what were the similarities and differences between social movements in small-scale polities and those that occurred in larger-scale chiefdoms and states? As we have already said, the biggest differences were in the social structural contexts. In egalitarian polities, authority was more diffuse but it still existed; and movements could define themselves as revised versions of authority such as occurred in the contest between the traditional shamans and the dream doctors at the time of the Ghost Dances. The diffusion of new ideas was constrained by modes of transportation and by simple technologies of communication in systems of small-scale polities. The new words, songs, and dances had to be carried on foot. This was a limitation on how far and fast ideas could spread, and the consequences of collective actions were more confined in space and operated more slowly in time. In multilingual situations, the signal-to-noise ratio was low and many of the original ideas were lost in translation. This could be seen in the mixing of the Ghost Dance ideas with other, earlier, cult forms in indigenous California. Indirect, down-the-line transmission of information also adds noise. Sign languages were also in use in indigenous North America (Davis 2010), but these were not good media for transmitting the subtleties of new cult ideas (Mooney 1965: 19). Obviously, the invention of writing and literacy increased the signal-to-noise ratio and allowed ideas to spread much faster and farther, not to mention the telegraph, the radio, and the internet (Massey 2005). New "religions of the book" transformed rituals of the spoken word into worship of the text.

Boat and caravan transportation facilitated the more rapid and distant spread of ideas. Institutions such as money and the law emerged in part as efforts to control social movements from below, but they could also be used by social movements. The same goes for economic and military organization. Bureaucracies and formal organizations were usually created to reproduce social orders, but social movements could also appropriate these inventions and use them to mobilize social change. The use of everyday institutions and organizations to launch movement campaigns acts as a recurring feature in the history of collective action (Almeida 2019b).

Revolutions in the Ancient World

Jack A. Goldstone (2014: chap. 4) devoted a chapter of his *Revolutions: A Very Short Introduction* to "Revolutions in the Ancient World." We have much better evidence regarding social movements once polities had developed the ability to record events in writing. The rise of primary or pristine states occurred in at least six, largely unconnected, world regions: Mesopotamia, Egypt, the Indus River Valley, the Yellow River Valley, the Andes, and Mesoamerica. These were dramatic instances of parallel sociocultural evolution in which somewhat similar conditions led to the emergence of larger settlements and larger polities. In all these cases, food production techniques—mainly irrigation—the domestication of animals, and the application of the plow, had developed to make possible the feeding of large numbers of people living close together in large cities. The cases of Cahokia and Mesopotamia previously discussed were both instances of this kind.

Research on scale changes in the population sizes of cities and territorial sizes of polities shows that around half of the cases of upward sweeps in these size indicators were the result of conquests by noncore marcher states (Inoue et al. 2016).[24] This confirms our hypothesis that core/periphery relations and uneven development are important for explaining the emergence of complexity and hierarchy in world-systems, but it also shows that a significant portion of upsweeps were not associated with the actions of noncore marcher states. We have developed a multilevel model (Chase-Dunn and Inoue 2019) that combines inter-polity dynamics with the "secular cycle" model developed by Turchin and Nefedov (2009). This model includes social movement mobilization and revolutions along with demographic and economic variables. Goldstone's chapter is useful because it suggests that his demographic model of state collapse (formalized by Turchin and Nefedov [2009] as a "secular cycle") can be applied to chiefdoms and early states as well as classical and modern ones. Goldstone's earliest example of revolution occurred during and after the reign of Pepy II, the last pharaoh of Old Kingdom Egypt whose reign ended during the twenty-second century BCE. The power of the central government was being challenged by regional nomes (elites) who were building large funerary monuments to themselves in what Van der Mieroop (2001: 81) describes as the "democratization of the hereafter" because the local leaders were presuming to join the pharaoh in the glorious next world. Social order was breaking down. The poor displayed little regard for those of rank.

Goldstone also describes the social movements and revolutions known from the classical Greek and Roman worlds (see also Collins and Manning 2016). Here, there is much more documentary and literary evidence describing the processes of social movements and revolutions. Goldstone recounts the unsuccessful struggles of the Gracchus brothers in Republican Rome to protect the interests of farmers against the acquisition of large plantations by slave-owning latifundistas (Brunt 1971). Peasants revolts, slave revolts, urban food riots, attacks by pastoral nomads, bandits, pirates, and internecine struggles among elites occurred repeatedly in the context of the rise and fall of regimes and of empires in the ancient and classical world-systems. Turchin (2003: chap. 9) notes the relevance of Ibn Khaldun's cyclical model of the rise and fall of regimes and the importance of changing levels of *asabiyah* (loyalty, solidarity, and group feeling) as old urban regimes became decadent and nomadic pastoralist polities emerged from the desert to conquer the old core and to form new urban regimes (see also Amin 1980; Chase-Dunn and Anderson 2005; Anderson 2019). War leaders from the noncore enjoyed greater support from their warriors than did the older, urbanized elites. This is an example of how collective action is an important dimension of struggle among competing elite factions.

The rise of the world religions during and after the Axial Age displays the interaction between social movements and forms of governance.[25] The term "world religions" in everyday discourse simply means organized religions with large numbers of adherents in the contemporary world. Here, we use it in a narrower sense to mean religions that combined universalistic claims with a proselyting mission that was expansive across kinship and language groups. World religions in this sense separated the moral order from kinship, allowing for, and encouraging, the inclusion of nonkin into the circle of protection. This was the expansion of human rights beyond the bounds of kinship and the expansion of what Peter Turchin (2016a) calls "Ultrasociety"—altruistic behavior among nonkin. Marvin Harris (1977) pointed to the frequency of ritual cannibalism practiced on enemies in systems of small-scale polities. In small systems, nonkin were not really humans. They are often enemy others that were not due any positive reciprocity (Sahlins 1972). The question of who the humans are and who are not the humans is important in all cultures. In small-scale polities, the distinction between "the people" and the non-people is usually a mixture of kinship relations and familiarity with a language. The moral order applies to the circle of "the people" and heavy othering sees the non-people as animals or enemies. When world-systems expanded, important debates occurred as to whether newly encountered peoples had souls or not. The rise of what we currently call humanity as a social construction was a long, slow, back and forth, uneven process that continues in the current struggle over citizenship (Hunt 2007; Joas 2013).

World religions locate great agency in the individual person even if it is only the right to declare obedience. To become a member of the moral order, a convert must confess and proclaim belief in the godhead. This is the act of an individual person. One's own action is required. Non-world religions usually tie membership to one's birth parents. Salvation is also a further democratization of the hereafter. Now the masses, too, may go to heaven or become enlightened.

Marvin Harris (1977) contended that the rise of world religions was functional for expanding empires because they included the conquered populations within the moral order of the conquerors. This proscribed cannibalism and reduced the amount of resistance mounted by the prospectively conquered population. The emperor will make you pay tribute and taxes, but he promises not to eat you. Most world religions began as social movements from the semiperiphery or the periphery (Bactria, Palestine) that were eventually adopted by the emperors. Prophets and charismatic leaders mobilized cadres who spread the word orally and with written documents. Sects and communities of believers were organized, eventually producing formally structured churches. Older institutions resisted, often repressing the new movements, but they continued to spread, in some cases becoming conquering armies and in other cases being adopted by kings and emperors. Some of the world religions were monotheistic, but others had no godhead at all except the path to enlightenment. Peter Turchin's (2016a: chap. 9) depiction of the rise of the world religions during the Axial Age focuses on the importance that these moral orders had for the construction of legitimate authority. The emperor was supposed to also obey the religious commandments, at least in theory. This gave the conquered an entitlement to membership in the moral community and provided a basis for claims against the authorities if they were violating the rules. Turchin mentions that monotheism puts God above the Emperor, but in China the same function was performed by the idea of the mandate of heaven in the absence of a daddy in the sky. These religions often included a large dollop of magical wishful thinking that gave hope to the downtrodden, such as the promise of immortality.

There is a well-developed and convincing literature on early Christianity as a social movement (Blasi 1988; Stark 1996; Mitchell, Young, and Bowie 2006). We have already mentioned the world-system setting of the origin of the movement. Christ and his followers emerged in a context of a powerful Roman colonialism in which the colonized peoples were faced with overwhelming force (Ellacuría 1985). The notions of individual salvation, rendering unto Caesar what is Caesar's, and concentration on the rewards of life after death were powerful medicine for those who faced a mighty Roman imperialism. Paul's mission to other colonized peoples and the delinking of salvation from ethnic origin was a recipe that allowed the movement to spread back to the poor peoples of the core. And eventually it was adopted by the Roman emperors themselves as a universalistic ideology that could serve as legitimation for a multi-ethnic empire. The prince of peace and salvation ironically proved to be a fine motivator for later imperial projects such as the reconquest of Spain from the Moors and the conquest of Mesoamerica and the Andes (Padden 1970). And as Cora DuBois ([1939] 2007: 116) has said, it also worked for the conquered as a way for them to survive psychologically and to adapt to a world in which their indigenous lifeways were coming to an end, recall the previous discussion of "revitalization movements" in this chapter.[26]

Another interesting characteristic of early Christianity is that it was primarily an identity movement. The early Christians did not propose to change the larger social or political order (Blasi 1988). They simply wanted the freedom to be allowed to have dinner together in Christ's name. Repression of the movement by authorities who saw this as a violation of religious propriety was a driving force in the formation of the early Christian communities.

Hinduism and Confucianism are not very proselytizing but they both spread successfully because they provided new justifications for hierarchy and state-formation. The spread of Hinduism to mainland and island Southeast Asia occurred because its notion of the god-king (deva-raja) provided a useful ideology for the centralization of state power in a context of smaller contending polities (Wheatley 1975). Confucianism provided a different justification for state power based on the notion of the mandate of heaven. It spread from its original heartland to the rest of China and Korea.

East Asia also saw periodic eruptions of popular heterodox religious movements. Mass heterodox movements are known to have been a recurrent feature of East Asian dynastic cycles (Anderson 2019). In the Han dynasty during a period of peasant landlessness, large numbers of poor people were drawn to worship the Queen Mother of the West who grew longevity peaches that, once eaten, made people immortal (Hill 2015). The Queen Mother lived in a mythical palace on a mountain somewhere in the West. This idea seems to have been present as early as the Shang dynasty, but recurrent eruptions of the worship of the Queen Mother corresponded with periods in which there were large numbers of landless peasants. The attraction of stressed masses to "pie in the sky when you die" reoccurs in world history. The White Lotus movement was another heterodox popular movement that first emerged in China during the eleventh century and became powerful during the Yuan dynasty. Ming dynasty founder Zhu Yuanzhang was an adept. It had ideological elements such as gender equality, vegetarianism, and egalitarianism that reappeared in the gigantic Taiping Rebellion in the middle of the nineteenth century (Spence 1996). The Taiping Rebellion was part of an interesting confluence in which the East Asian dynastic cycle was becoming entwined with the world revolutions of the West. The Taiping founder and leader, Hong Xiuquan, read a pamphlet about God and Jesus that a Christian missionary from Tennessee had translated into Chinese. After repeatedly failing the imperial examination, Hong had a dream in which it was revealed to him that he was Jesus's brother. The Taiping movement morphed into a large army that conquered Nanjing. As many as thirty million people died as the Qing dynasty fought for decades to extinguish this rebellion.[27]

As we have mentioned, Turchin and Nefedov (2009) showed that secular cycles linking demographic waves with dynastic fall and rise operated within many agrarian polities. Peter Turchin (2016b) has convincingly applied this model to cycles of political discord within the United States in the nineteenth and twentieth centuries. But the secular cycle model is operationalized in a way that treats inter-polity relations (trade and warfare) as exogenous to the processes of demographic growth, elite competition, and state collapse. The multilevel model suggested in this chapter incorporates inter-polity interactions with the secular cycle operating within polities to provide a structural demographic framework for explaining the evolution of world-systems (Inoue and Chase-Dunn 2019).

Ho-Fung Hung (2011) found that proactive movements that confronted the local authorities and engaged with the imperial center, often led by local gentry, emerged in waves during the Qing dynasty. He contends that in China there was a cycle in which proactive movements and reactive movements oscillated. He claims that the Western emergence of proactive movements focusing on popular sovereignty (rather than filial loyalty) was unidirectional, leading to the establishment of parliamentary democracies. This simplifies the contentious nature of the emergence of "Centrist Liberalism" in the West (Wallerstein 2011) allowing teleology back in (see chapter 5). The world history of the first half of twentieth century (two world wars, Bolshevism, and Fascism) surely contradicts the idea of a smooth transition to capitalist democracy in which rational proactive movements fine-tuned the ability of human institutions to meet the needs of the people. Identity movements, religious fundamentalisms and populist nationalisms in the twenty-first century still strongly contest the notion that the modern world-system has risen above reactive social movements.

David Zhang and Harry Lee and their colleagues are pioneers in the long-range study of East Asian conflict and its relationships with climate change (Zhang et al. 2006, 2007, 2015; Lee 2018). They built on the earlier research of J. S. Lee (1931, 1933), who studied the relationship between "internal wars" and climate change in China.

The Zhang et al. (2007) findings for the number of rebellions from 1000 to 1911, counting events per decade, is shown in figure 1.1. Zhang et al. (2007: 405) said, "We . . . categorized wars on the basis of types of participants, particularly the leaders of the two sides in the armed conflicts, as either 'rebellion' or 'others' (state and tribal wars)." The rebellions were predominantly peasant uprisings induced by famine and heavy taxation, since farmers were always the first to suffer from declining agricultural production. The three outstanding peaks of warfare were dominated by peasant uprisings. Wherever they occurred rebellions are always significantly correlated with temperature."

In an appendix to their 2015 article, Zhang et al. (2015) say more about the relationship between climate and conflict:

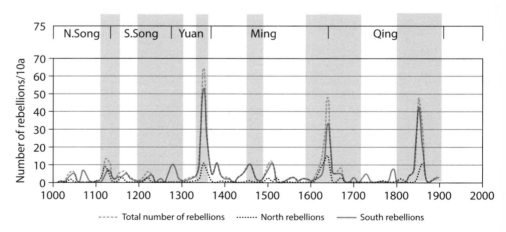

FIGURE 1.1   Rebellions in China, 1000 CE to 1911 CE. (The gray shaded areas represent colder phases.) *Adapted by permission from Springer Nature: Zhang et al. 2007: 406, fig. 2d.*

What caused the cycles of nomadic invasion and retreat or the expansion and shrinkage of agriculturist empires? Lee (1933) conjectured that the power of nature was the answer. Some scholars have found strong statistical relationships between climate change, war, population and dynastic cycles in Eurasia and the world by using high-resolution temperature reconstructions and fine-grained historical datasets (Zhang *et al.*, 2006, 2007a, 2008; Tol & Wagner, 2010).

This study decomposed history into different time domains and found that the multicentennial geopolitical changes are associated with climate change, but the short-term changes (less than 200 years) do not exhibit any rhythmic pattern. This implies that those short-term geopolitical changes might be associated with social and political changes that are non-linear and irregular (no continuous frequency). The results indicate that a complex system may be controlled by different factors at different spatial and temporal scales (O'Neill *et al.*, 1989; Norton & Ulanowicz, 1992). . . . When a selected factor has a characteristic time-scale which is adequate to the scale of the considered process, while all the other factors have significantly different time-scales, we can consider the selected factor as the most evident one that controls the process (Korotayev *et al.*, 2006). In this study, we could infer our ecogeopolitical hypothesis by the fact that the long-term geopolitical and precipitation frequencies were strongly coherent in China at the multicentennial scale, but the short-term geopolitical changes, even if they were of lower magnitude, might have been controlled by other, unknown factors. (Zhang et al. 2015: appx. S3, 15–16)

The Zhang et al. (2007) paper shows that there have been cyclical waves of rebellions in China from 1000 until 1911. This is consistent with the hypothesis that the East Asian world-system experienced world revolutions, but further research is needed to include rebellions in Korea, Japan, and Southeast Asia, and to examine the relationships between these and dynastic cycles and upsweeps of polity and settlement sizes. The question as to why cold spells should cause increased conflict is complicated. Regarding peasant rebellions Zhang et al. argue that it was declining agricultural production, famine, and taxation that stimulated rebellions. The studies by Zhang et al. usually found a two or three-decade lag between cold peaks and the onset of conflict. Harry Lee (2018) suggests that a bad climate and decline in agricultural productivity often sets the stage for natural disasters (floods) or sociocultural catastrophes (famines, epidemics) that serve as the proximate causes of the outbreak of rebellions.

In the modern world-system, there has been a spiral of interaction between world revolutions and the evolution of global governance. As we have said in the introduction, world revolutions are periods in world history in which many rebellions break out across the world-system, often unconnected with one another, but known, and responded to, by imperial authorities. Since the Protestant Reformation, such constellations of rebellions and social movements have played an important role in the evolution of global governance because the powers that can best handle collective behavior challenges are the ones who succeed in competition with competing elites. It is possible that similar phenomena existed in prehistorical and historical world-systems. Oscillations in the expansion of trade networks; the rise and fall of chiefdoms, states and empires; and increasing synchronization of trade and political cycles are likely to have been related to the waves of social movements and rebellions that occurred in the same time periods among polities that were interacting with one another (world revolutions). This hypothesis is a subject for future research. Our claim that social movements have been an important cause of longrun cycles and upsweeps of complexity and hierarchy requires attention to what has been happening more recently. We next turn to movements in the context of the "globalization project"—those battling capitalist neoliberalism in the twentieth and twenty-first centuries in the Global North and the Global South.

TWO

Resistance to Neoliberalism in the Global North and South
National and Local Dynamics

The previous chapter focused on collective action and social movement–type activity in the premodern world—a period lacking in scholarship on global change and social movements. In this chapter, we will focus on global economic change and social movements in the past century and the current one. We will once again observe the role of threats, harms, and disruptions of ordinary people's daily lives as catalysts to engage in collective mobilization. More precisely, we emphasize the global transition from state-led development to neoliberalism as creating the cleavages for large-scale campaigns of resistance at the local and national levels in most world regions.

Even in the contemporary neoliberal period characterized by hyper-individualism, the intensification of the culture of mass consumerism, and the substantial downsizing of the welfare state, social movement activities based on social solidarity continue to be sustained around the globe. Given this unfavorable political and economic context, such upsurges in collective action would seem difficult to explain. We argue here for a historically rooted understanding of the contributions of the prior period of state-led development in shaping the possibilities for collective action campaigns resisting neoliberalism in the twenty-first century. The state expansion of health/medical care, mass schooling, transportation, and economic infrastructure in the mid-twentieth century conditions the level of resistance in the twenty-first century to new forms of economic liberalization. The period of state-led development from the 1930s to the 1980s is unprecedented in world history for the rapid rates of infrastructural development that occurred throughout the world (Cohn and Blumberg 2016). This newly formed infrastructure provides the organizational bases for the collective opposition to neoliberalism in the current century.

Threat and Organizational Infrastructures
in Social Movement Theory

Social movement theory is increasingly giving more attention to the negative conditions and grievances driving popular collective action (Almeida 2018). Following critiques of relative deprivation theories of mobilization, scholars downplayed the role of these negative conditions or threats on catalyzing social mobilization in the late twentieth century. The excessive psychological focus of the various strains of relative deprivation theory on individual states of frustration and aggression left students of social movements dissatisfied with its core causal explanations (Buechler 2011). Over the past two decades, analysts of social movements have increasingly examined threats as a *social* process that affect large groups of people in a similar fashion (Van Dyke and Soule 2002; Dixon and Martin 2012; Simmons 2014; Snow and Owens 2014).

Two particular threats that have drawn large numbers of groups into the streets include (1) state-attributed economic problems and (2) the erosion of rights (Almeida 2003).[1] State-attributed economic problems involve economic shifts that are blamed on the state such as loss of land, consumer price hikes, subsidy cuts to basic goods and services, wage arrears, and widespread unemployment. Erosion of rights involves the loss of citizen rights that were previously established via constitutional means or more informally through social contracts between the state and civil society (Oxhorn 2011; Rubin 2012). With the ascendancy of neoliberal regimes around the globe, the two threats of state-attributed economic problems and the erosion of rights have combined into the threat of weakening *social citizenship rights*. Social citizenship rights incorporate state protections from the market as well as the provision of social benefits (Marshall 1950; Meyer and Evans 2014). Beyond religious and ethnic conflicts, as well as struggles against authoritarian states, the deterioration in social citizenship rights drives some of the largest mobilizations and protest campaigns in the era of globalization—from Spain, Greece, and Portugal in southern Europe to the wider Global South in Ecuador, Chile, Colombia, Honduras, Iraq, Iran, Lebanon, Nigeria and South Africa.

Figures 2.1 and 2.2 below illustrate the struggles across the planet against neoliberalism. Figure 2.1 displays the yearly number of massive

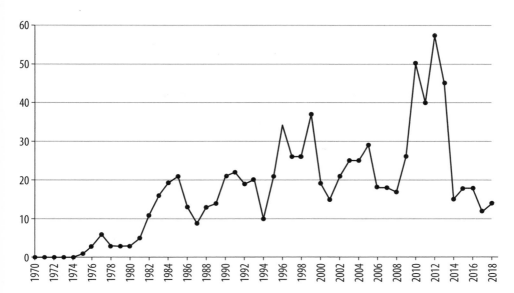

FIGURE 2.1 Worldwide Massive Outbreaks of Popular Unrest against Neoliberalism, 1970–2018 (nationwide campaign and/or more than 100,000 participants). *Lexis Nexis Academic Universe 1970–2018; Social Conflict Analysis Database (SCAD); and the Urban Social Disorder (USD) Dataset (Urdal and Hoelscher 2012).*

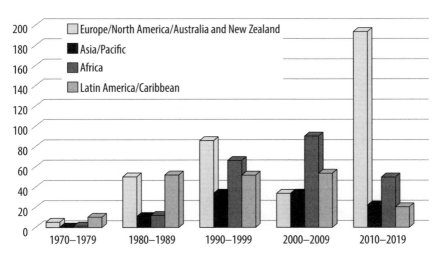

FIGURE 2.2 Massive Protest Campaigns against Neoliberalism by World Region, 1970–2019 (nationwide event and/or more than 100,000 participants). *Lexis Nexis Academic Universe 1970–2019; Social Conflict Analysis Database (SCAD); and the Urban Social Disorder (USD) Dataset (Urdal and Hoelscher 2012).*

outbreaks of unrest against neoliberal policies (nationwide protests and/or protest events with over 100,000 participants). Neoliberal policies were defined as protests against austerity, price hikes, privatization, labor flexibility, and free trade agreements. The periodization runs from 1970 through 2018. Figure 2.2 displays the same data through the first half of 2019 broken down by world region and decade.[2]

Another challenge for scholarship that seeks to demonstrate the mobilizing role of threats to social citizenship centers on specifying the organizational conditions under which negative economic environments activate collective action. Preexisting social networks and organizations are critical for determining if threats will convert into large-scale movement campaigns as well as the pace of mobilization (Morris 1984; Edwards, McCarthy, and Mataic 2018). Because of the ubiquitous nature of threats and the lack of mobilization in most times and places (McAdam and Boudet 2012), the types of organizational contexts promoting threat-induced mobilization need to be clearly established. In this case, we are primarily interested in the organizational mechanisms leading to mobilization from the threat of eroding social citizenship rights. Past cross-national and national research on resistance to neoliberalism has highlighted the roles of labor unions, levels of industrialization/GNP, urbanization, and public sector employment as providing the resources and organizations for collective resistance (Walton and Ragin 1990; Walton and Shefner 1994; Abouharb and Cingranelli 2007; Anner 2011; Ortiz and Béjar 2013; Beissinger and Sasse 2014). Other studies emphasize the part played by community and state infrastructures such as NGOs (Gallo-Cruz 2019), oppositional political parties, public universities, state administrative offices, and highways in sustaining opposition to contracting social citizenship rights (Walton and Seddon 1994; Roberts and Portes 2006; Almeida 2012, 2014).

At the national and subnational levels, scholars are beginning to recognize that, while there are homogenizing impacts of globalization at the macro level (Centeno and Cohen 2010), such as world society scripts adopted by individual nation-states pushing in the direction of isomorphism, global intrusions within nations and localities are highly uneven (Sassen 2008; Lobao 2016) and generate different kinds of responses depending upon local conditions. The greatest body of literature on this theme comes from studies of national-level collective resistance to the family of policies related to market-driven globalization: economic austerity, structural adjustment, privatization, labor flexibility, and free trade (Roberts 2008). In both the Global North and the Global South, the level of mobilized opposition to neoliberal policies has varied across countries, within countries, and over time.

Beginning as a global trend in the 1980s, the debt crisis led to dozens of uprisings across the developing world (see figure 2.2). The rebellions were tied to the structural adjustment policies implemented by governments in the Global South as a condition for loan rescheduling with international financial institutions (Babb 2005, 2013). Walton and Seddon (1994) have convincingly demonstrated how the third-world debt crisis placed dozens of countries in similar economic circumstances, with common structural adjustment packages negotiated. By 1984, sixty-five debt reschedulings had taken place. With most developing countries spending over a decade of structural adjustment under either International Monetary Fund (IMF) or World Bank conditionality (Abouharb and Cingranelli 2007), a new wave of austerity and anti-neoliberal protests emerged in the late 1990s and 2000s over privatization, free trade, and other economic liberalization policies (Eckstein 2002; Silva 2009). In the aftermath of the 2008–2009 Great Recession, the Global North also experienced a massive wave of austerity protests (Castells 2012; della Porta 2015; Kanellopoulos et al. 2017). Figures 2.1 and 2.2 demonstrate the sharp rise in austerity protests in the aftermath of the Global Recession in the 2010s.

These economic-based protests offer an interesting challenge to extant social movement studies by their relationship to global dynamics, by negative circumstances triggering their emergence, and by the intermediary conditions of local-level contexts. The debt crisis, the institutionalization of neoliberalism, and the Great Recession all operated on a worldwide scale (Cohn and Hooks 2015; Kentikelenis and Babb 2019). Economic protests are largely generated by economic threats to social citizenship rights in terms of declining standards of living and livelihoods for middle- and lower-economic strata groups (Almeida 2018). Recent scholarship attempts to specify when economic threats tied to global economic liberalization will lead to national- and local-level collective action (as opposed to focusing exclusively on the positive conditions associated with political opportunities) (Caren et al. 2017). Below, basic expansions of the welfare state and economic infrastructure are linked to the key organizational foundations and motivations for local and national resistance to neoliberalism and the loss of social citizen rights over the past four decades.

State Building and Economic Development in the Global South

Chapter 1 traced social movement–type activity back to early human settlements. Charles Tilly (1984) has demonstrated the coevolution of the modern national social movement with the expansion of states and national markets in nineteenth-century Europe. Centralizing states within bounded territories and governed by parliaments channeled popular mobilization onto a national scale as grievances and demands implicated citizens beyond locally-based issues. The state at the national level became the final arbiter of policy affecting the citizenry. Related processes of the spread of mass literacy and the printing press also made the possibility for the awareness of common interests across wide geographical spaces (Tarrow 2011).

Capitalist development in the nineteenth century also facilitated the rise of trade unions and political parties in cities that would extend to other regions (Tilly and Wood 2012). Developing large-scale manufacturing along with the centralization of the working class made possible the capacity to sustain social movement–type mobilization for improved working and housing conditions along with other economic, political and social rights that European states increasingly granted (Oxhorn 2003; Mann 2013). Hence, national social movements arose in Europe during the initial phases of the emergence of industrial capitalist development. A somewhat similar process would unfold in the twentieth century in the developing world.

Development State Breakthrough in the Mid-Twentieth Century

In economic terms, the post–World War II period in the Global South is known as the era of "state-led development" (SLD), which provided a comprehensive plan for developing economies to achieve rapid industrialization and modernization (Kohli 2004). This preglobalization, state-led development era (also referred to as Fordist-Keynesianism) runs roughly from the 1930s to the early 1980s (Robinson 2014). The development strategy centered on a massive expansion of the state infrastructure and reinvesting primary agricultural and resource extraction industry surpluses into manufacturing to reach a stage of self-propelled economic growth (Rostow 1960) via import substitution industrialization (Kay 1989). To back this process up, governments in the developing world, with the aid of international economic assistance and expertise, expanded both the social and economic infrastructure of public education, state administration, transportation, health care, banking, telecommunications, and electrical power.

Segura-Ubiergo (2007: 1) defines the welfare state as, "a repertoire of state-led policies aimed at securing a minimum of welfare to its citizens," including public investments in health and mass education. The expansion of the state infrastructure coincided with the creation of a new belief system within civil society and a set of expectations called *social citizenship* (Marshall 1950). Social citizenship has been defined as "the rights and duties associated with the provision of benefits and services designed to meet social needs and enhance capabilities, and also guarantee the resources necessary to finance them" (Taylor-Gooby 2009: 4). While these social citizenship rights first extended in Europe in the nineteenth century with the legalization of labor unions, associations, and access to vital social services, these benefits were greatly expanded in the aftermath of the Great Depression and World Word II with the advent of modern welfare states across the advanced capitalist world (Mann 2013), followed by attempts at emulation throughout the Global South (Cohn 2012). The benefits of social citizenship were also unevenly divided in racialized states (Fox 2012).

This extension of social citizenship to populations in the developing world during state-led development occurred under a variety of regimes including military-controlled governments, a variety of authoritarian and corporatist states, populist rule, and democracies (Oxhorn 2011). The new social programs and subsidies of the developmental regime became engrained as the "moral economy" of the period and as an implicit "bargain" between the popular classes and the state—an exchange of political loyalty for social and material support (Walton and Seddon 1994: 46–48). As in the Global North, not all social sectors benefited equally from the extension of social welfare in the Global South; indigenous peoples, women, ethnic minorities, and rural and informal sector workers received less access to education, health care, and social services. Nonetheless, even subaltern groups were relatively better off with the expansion of social citizenship and economic infrastructures in terms of life chances and mobilization potential than in prior decades.[3]

As stated above, the modern welfare state systems of the advanced capitalist nations were emulated throughout the developing world from the 1930s through the 1980s. While basic public health systems were established early in the twentieth century, in the post–World War II period, there was rapid growth in building public hospitals and clinics in the Global South. This growth in health-care access also corresponded with the establishment of social security systems for the urban working and middle classes in the same time period that often combined a retirement package with health and medical care coverage (Mesa-Lago 2008). In the current period of market-driven globalization (1980s–2010s), there has been a reversal in state commitments to social insurance and health-care coverage (Mesa-Lago 2007).

In addition to public health, institutions were greatly expanded in mass education in the state-led development era. Developing countries invested heavily in constructing primary and secondary schools as well as universities. Beginning in 1940, primary school enrollments reached a logarithmic rate of growth around the world which drove up high school and university enrollments in the following decades (D. Baker 2014). By the 1970s, most municipalities in Latin America (even in remote rural regions) had at least a primary school. Schofer and Meyer (2005) have documented the global increases in university enrollments, with an especially accelerated rise in Africa, Asia, and Latin America in the 1960s and 1970s. While the largest increases in university enrollments occurred in the central public universities in the capital cities of the developing world, governments also established branch and regional campuses of the public university system, extending access to higher education to more distant provinces. Countries with higher concentrations of university-educated publics have been associated with the emergence of nonviolent movements in the world-system (Gallo-Cruz 2019).

At the same time as states extended health and educational opportunities across their national territories, developmentalist regimes also subsidized consumption via price controls, on basic grains, transportation, medicine, and even agricultural inputs for small land holders in the rural sector. The trend was similar for Africa, Asia, Latin America, and Eastern Europe. Government policies of price controls and subsidies for basic goods served as an integral part of the social citizenship compact between the state and civil society.

Outside of health, education, consumer and farmer subsidies, and social services, the development state also invested heavily in basic economic infrastructure to propel economic growth. This included the construction of hydroelectric power and nationally integrated energy grids, national highway networks, and water and sewage systems. Taken together, all of these development efforts in the public and economic spheres sought to support the partial industrialization of the third world, and largely succeeded in the semiperipheral world (Chase-Dunn et al. 2015). These state investments in economic development appeared to pay off as developing countries slowly began to increase their share of global industrial output and share of exports in manufacturing between 1960 and 1980 (Dicken 1986). By the late 1970s, state-owned enterprises (SOEs) accounted for 25 to 30 percent of total domestic investment in Africa, Asia, and Latin America (A. Baker 2014). This unprecedented expansion of the state social and economic infrastructure eventuated in a major outcome beyond the unprecedented rates of urbanization, economic growth, industrial output, and upward mobility—it also changed the nature of collective action throughout the developing world.

State-Led Development as Organizational Capacity for Mobilization

The process of state-directed development created a large jump in scale in the potential for popular collective action across a broad geographic space. Newly industrialized countries such as China, South Korea, South Africa, India, Egypt, Turkey, Nigeria, Argentina, Brazil, Chile, Uruguay, and Mexico grouped large numbers of manufacturing workers in similar circumstances. In an even broader group of peripheral states, the expansion of public administration, schools, health systems, and transportation networks raised the mobilization potential for the population in general and specific social sectors (Klandermans 1997). By the 1960s and 1970s, state and economic infrastructures reached such a degree of development whereby mass movements of teachers, industrial workers, students, peasants, civil servants, and squatters could sustain protest campaigns to push for an even greater extension of social citizenship rights via higher wages, social welfare benefits, and basic protections of livelihood. These struggles often evolved into struggles over *political* citizenship rights in contexts of authoritarian and semi-authoritarian rule (Markoff 2015b).

The process of constructing national highway networks during the state-led development era made it possible for aggrieved groups to travel to provincial and national capitals to present their claims to state and elected officials, once again increasing the scale of collective action from the local level to the regional and national levels. The building of schools and investing in mass education resulted in national associations of school teachers often serving as the largest labor-based organization in many developing countries (Silver 2003). As high school and university enrollments accelerated in the 1960s and 1970s, an increase in youth- and student-based movements also could be observed throughout the Global South.

Most importantly for our purposes here, the rapid and extensive expansion of social welfare and economic infrastructure not only raised the scale of mobilization in the state-led development period, but also deposited organizational assets that would be the most important basis to launch collective action campaigns against market-driven globalization in the twenty-first century. A set of bureaucratic practices and structures were put in place to administer the expanded infrastructure that persisted into the neoliberal era (Evans and Rauch 1999).[4] Once established, these infrastructures provided a fungible resource to be used by collective actors for a variety of purposes. The collective opposition to the erosion of social citizenship rights has been one of the primary uses of this fungible development state infrastructure. The largest mobilizations against neoliberalism in Eastern Europe, Africa, Asia, and Latin America are led by state sector organizations or sectors that greatly benefited from the expansion of social welfare in previous decades (see case of Costa Rica below). For example, in one empirical study of 281 anti-neoliberal protest campaigns in Latin America, public employees, students, and school teachers were among the specific social sectors with the highest rates of participation

(Almeida 2007). In Africa, much of the large-scale organized opposition to free market reforms derived from state sector labor associations (including school teachers) in alliance with community groups.[5]

Threats to Social Citizenship as Catalyst of the Opposition to Neoliberalism

It is largely the actual loss and erosion in social citizenship rights that provide the threat incentives for collective mobilization over the past four decades. The crucial turning point begins with the breakdown of Keynesianism in the 1970s and third-world debt crisis of the early 1980s (Robinson 2004). The inability of indebted states in the Global South to make payments on foreign debt to banks and governments in the Global North led to the intervention of the World Bank and the IMF. The World Bank and the IMF renegotiated foreign debt with third-world states in exchange for structural adjustment agreements that forced governments to enact austerity policies, labor flexibility laws, privatization, and open their economies to greater foreign investment (Vreeland 2003). Structural adjustment policies persisted through the 2000s across the developing world, eventuating in the degradation of already feeble welfare states (Shefner and Blad 2020). Even in Africa, structural adjustment policies beginning in the 1980s ended the "postcolonial social contract" (Carmody in Barchesi 2011: 17).

Figure 2.3 illustrates the scope of structural adjustment in the Global South. The map shows the number of years individual countries were under an IMF or World Bank structural adjustment agreement between 1981 and 2008. As the map depicts, the regions of Africa, Latin America, and Central-Eastern Europe were under structural adjustment programs for over ten years during the global transition to neoliberalism—much of the Global South. It is precisely these world regions that reported the highest outbreaks of anti-neoliberal protest in the earlier years of the debt crisis through early 1992, with Latin America alone accounting for 62 percent of all events (91 out of 146) documented in the Global South (Walton and Seddon 1994: 39–40). The graphic in Figure 2.3 also assists in explaining the similarity in the grievances in terms of the erosion in social citizenship as dozens of countries in several world regions found

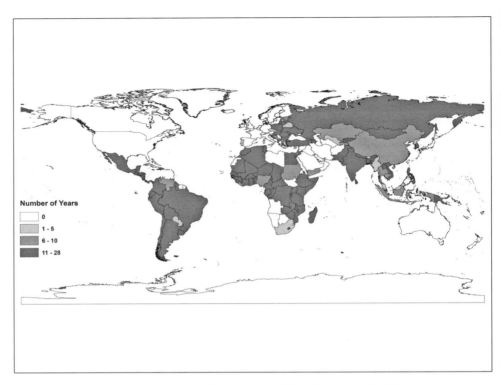

FIGURE 2.3 Number of Years under IMF–World Bank Structural Adjustment between 1981 and 2008. *The authors appreciate the updated data on structural adjustment agreements provided by Abouharb and published in Abouharb and Duchesne (2019).*

themselves under similar circumstances of austerity and social welfare retrenchment. Since 2008, the external debt has continued to increase in the regions experiencing long-term structural adjustment, while Southern European nations have also come under structural agreements.

Figure 2.4 conceptually summarizes the framework presented above. The expansion of the development state infrastructure and its policies of redistribution (1940–1980) created both the organizational capacity for widespread popular mobilization on a national scale and a compact of social citizenship between civil society and the state. The massive and unprecedented expansion of health care, education, public administration, and economic and transportation infrastructures placed large numbers of people in similar

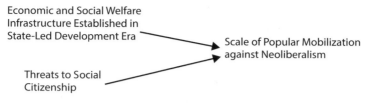

FIGURE 2.4   Resistance to Neoliberalism

circumstances and provided them fungible assets to collectively launch social movement campaigns to expand the welfare state beyond the initial de-
sires of economic elites and populist power holders. When regimes in the
Global South transitioned to a neoliberal development trajectory in the
1980s and 1990s, the social citizenship gains of the previous four decades
came under threat acting as the central catalyst for mass mobilization.

The more extensive the state infrastructure created during the state-
led development era, the greater the capacity for civil society to mount
sustained campaigns of protest in the neoliberal period. In other words,
greater resistance to neoliberal reforms should be observed in societies
with larger state infrastructures. In addition, we can take a "fractal" approach and observe these same processes at lower levels of social organization (Goldstone 1991: 36). Hence, at the subnational level, by extension, localities with greater levels of state infrastructure should experience
higher levels of popular resistance to neoliberal reforms than communities that lack the relative presence of state infrastructural properties (Almeida 2012). In summary, there is variation across time, countries, and
within countries in the level and frequency of collective opposition to
market-driven globalization.

The Pattern of Mass Mobilization against the Neoliberal Model of Development

In this section, we outline the general parameters of the opposition to neoliberalism based on the loss of social citizenship rights and using the structures of the development state to mount the social movement campaigns to
protect civil society from the erosion of these rights. We draw on documentary evidence focusing largely on the Global South. The examples include

the major types of anti-neoliberal struggles, including food price hikes, anti-austerity, privatization, and free trade. We also demonstrate the perspective down to local level variations in resistance with the case of Costa Rica.

The expansion of administrative, welfare, and economic infrastructures under state-led development deposited a persistent set of social and physical structures that could be appropriated for collective action, even decades after they were first established. In particular, universities, public schools, the health-care sector, the public sector in general, and the national transportation network have played critical roles in generating and sustaining mass opposition to the neoliberal development model (Leal 2020). These portions of the state sector provide the sentiment pools for mass mobilization to materialize, increasingly with the use of social media, especially via Facebook, Instagram, Twitter, Telegram, and increasingly WhatsApp.

Just the scale of Costa Rica's "tropical welfare state" (Edelman 1999) has made it the site of multiple mass campaigns against neoliberalism. This case is especially noteworthy given the history of mass mobilization and revolution in the surrounding Central American countries, while the relatively tranquil and politically stable country of Costa Rica has witnessed many of the largest campaigns against neoliberal reforms. The case fits our model with the country benefiting from arguably some of the most extensive social citizenship rights granted in the developing world after World War II, including near universal health-care coverage and affordable access to public utilities (e.g., electricity and telecommunications) and public education.

Costa Rica also was one of the first developing countries to default on its foreign debt in the early 1980s. Between 1980 and 2019, several mass movements surfaced to confront the transition to the neoliberal development model. In 1983, a multi-sectoral movement arose to defend the public from an IMF-imposed price hike on consumer electricity. The campaign was clearly marked by state-led development infrastructures. The anti-IMF price mobilizations were based at the community level by the National Directorate for Community Development (DINADECO) organizations created by the developmental state in the late 1960s to support local economic development projects (Alvarenga Venutolo 2005). The key tactic of the movement was to blockade national highways (which were constructed between the 1950s and 1970s to integrate the country for national economic development). In the end, the mobilization succeeded in turning back the price hike and protecting consumers. The other major mass movement in the 1980s in Costa Rica involved small farmers that lost access to agricultural subsidies after two major structural adjustment agreements. The farmers also used the blockades of national highways to try and defend their access to state resources to sustain their rural economic livelihoods (Edelman 1999).

This trend of using the state-led development infrastructure to protect civil society from market reforms in Costa Rica continued in the 1990s and 2000s. In 1995, a protest campaign erupted over a third major structural adjustment agreement between the Costa Rican government and the IMF and World Bank. This structural adjustment loan focused on eroding the pension system for public educators and staff as well as a series of privatizations in the public sector and price increases on basic consumption items. These reforms mobilized the largest demonstrations in decades, reaching up to 100,000 participants. The backbone of the movement was the teachers' associations, numbering up to 40,000 members, and they coordinated about two-thirds of all protest events in the campaign (Almeida 2014). Public universities (staff and students) and public sector labor unions also participated in large numbers in the mobilizations.

In 2000, an even larger protest campaign erupted in Costa Rica over the threat of privatizing electricity and telecommunications in a single piece of legislation. With some of the most extensive and low-cost utility access in Latin America, a solid coalition of state sector workers, high school and university students held dozens of roadblocks and mass marches to turn back the privatization. The territorial variation of the campaign also followed a pattern of the distribution of the state infrastructure with protest events more likely occurring in localities with public universities and transected by one of the country's main national highways (Almeida 2012). Just a few years later, the largest demonstrations broke out in modern Costa Rican history against the Central American Free Trade Agreement (CAFTA) (between 2004–2007), with health-care workers, university staff and students, public sector workers, and school teachers leading over 600 marches and strikes, some actions reaching up to 150,000 participants at a time (Almeida 2014). The opposition to CAFTA centered on the concern that the free trade treaty would greatly debilitate the welfare state if implemented, including the outsourcing of medical care, utilities, and insurance (Raventós 2018). These social sectors resurfaced as the vanguard against an IMF fiscal reform in 2018.

In Argentina, Auyero (2002) contends that the explosion of social unrest in the late 1990s and early 2000s against neoliberal development was rooted in the privatization of state firms and the decentralization of public health care and public education. In a brief ten-year period between 1989 and 1999, employment in the largest state infrastructure entities (water/aqueducts, petroleum, electricity, and telecommunications) dropped from 500,000 to 75,000 workers (Auyero 2002: 29). Dismissed state employees formed the core of the unemployed workers' movement—one of the largest social movements in Latin American in the late 1990s and early 2000s (Svampa and Pereyra 2009). The unemployed workers (called "piqueteros") central tactic of blockading the national highway infrastructure surpassed the number of labor strikes (Rossi 2017).

Mass strikes and protests in health care by hospital staff and school teachers in public education were based on devolving these social citizenship services and rights from the federal level to the provinces where local government resources proved inadequate to maintain investment in these critical social institutions. The teachers' labor confederation (CTERA) formed the first national protest movement in neoliberal Argentina by erecting a massive tent city in the capital in 1997. The teachers, health workers, and the state employees' association (ATE) would combine to form the most powerful labor confederation in Argentina in the 1990s and 2000s (and arguably the most combative in Latin America), the Congress of Argentine Workers (CTA), capable of holding the largest mobilizations and general strikes (Silva 2009).

With the return of a pro-neoliberal government in 2016 and new structural adjustment agreements with the IMF, including a $57 billion structural adjustment loan signed in 2018, the largest in IMF history with any country, Argentina's labor confederations and civic organizations launched multiple days of action and general strikes, with public sector unions and teachers playing a leading role. The return of neoliberalism in Brazil in 2017, with the ouster of the Workers Party from the executive branch, labor unions, university students, and their allies also unleashed several national days of protest between 2017 and 2019 against free market reforms by the Bolsonaro and Temer administrations. Even under the Workers' Party administration in 2013, the largest mobilizations since the early 1990s broke out in dozens of cities over rising transportation prices led by the Movimento Passe Livre (Free Fare Movement)—reaching over two million protest participants. Then, in late 2019, Ecuador, Chile, and Colombia exploded in the largest outbursts of nonviolent unrest in several decades over fuel and transport price increases enacted by the executive branch (and pension reforms, education budgets, and privatizations). All three neoliberal governments declared special states of emergency, while in Chile and Ecuador state security forces arrested thousands and killed over a dozen demonstrators in failed attempts to quell the unrest (Somma et al. 2020).

Across Central America in the 1990s and 2000s, as the region transitioned to a neoliberal model of development, the largest battles over any grievance involved privatization and free trade (Haglund 2010; Spalding 2014). In Panama, Honduras, and El Salvador, thousands of doctors and health-care staff joined in coalition with university students, teachers, and workers to prevent the privatization of the health and social security systems between 1999 and 2019. Mobilizations against health-care privatization most often occurred in communities with state administrative offices, highways, universities, and health clinics and hospitals (Almeida 2014). Besides Costa Rica, major mobilizations against CAFTA surfaced in Guatemala and El Salvador, with school teachers playing a decisive role in mobilizing throughout the national territory in Guatemala. Also in Guatemala, public health and medical workers have served as a key ally with Mayan communities in mobilizing against electricity price hikes in the 2010s. School teacher and health-care labor associations have also led anti-neoliberal campaigns in Honduras and Nicaragua, coordinating as the vanguard within multi-sectoral organizations such as the Coordinador Nacional de Resistencia Popular (CNRP), the Frente Nacional de Resistencia Popular, and the Plataforma para la Defensa de la Salud y Educación in Honduras (Sosa 2013; Frank 2018; Sosa and Almeida 2019) and the Coordinadora Social in Nicaragua. Indeed, in one analysis of the anti-neoliberal protest wave against structural adjustment in Honduras between 2001 and 2005, Sosa (2013: 152) found that public health workers and school teachers participated in protest events more than any other social sector by a wide margin (out of a list of fourteen protesting sectors).

In the 2010s, mounting environmental crisis and new rounds of resource extraction combined with earlier neoliberal policies to produce explosive protests in Central America. In 2018, the Nicaraguan state faced a series of campaigns of fierce resistance to mining, construction of a trans-oceanic canal, and permissive fires in protected tropical forests. The protests culminated when the government implemented an IMF-recommended reform to the social security and pension system that reduced benefits. The neoliberal reforms initiated a massive uprising led by students in the public universities. By 2019, over three hundred had been killed by security forces and para-military groups, while an additional eight hundred Nicaraguans were reported as political prisoners. At the height of the campaign between April and September of 2018, over two thousand anti-government protest events were documented (Cabrales 2019). The citizens of El Salvador also continued to mobilize against water privatization in 2017 and 2018 with large-scale campaigns supported by political parties, public universities and youth, NGOs, environmentalists, the church, and the public water administration workers. In 2019, Honduran medical and health workers, along with school teachers and students, unleashed a massive, nationwide, three-month-long mobilization against a government attempt to simultaneously privatize health care and education (Sosa and Almeida 2019).

Mexico has also followed a similar path of major mobilizations against neoliberalism. The resistance commenced with the formation of the Frente Nacional en Defensa del Salario, contra la Austeridad y la Carestía (FNDSCAC) in 1982 and two national days of protest in 1983 and 1984 against austerity measures. An October 1983 national day of protest against economic austerity involved over one million people mobilizing over 120 simultaneous actions throughout the country. A similar day of mass action occurred in June of 1984. Resistance to neoliberalism continued in the 1990s with the 1994 Zapatista rebellion against the North American Free Trade Agreement (NAFTA) that sustained an indigenous people's cycle of protest in Chiapas for over a decade (Inclán 2018). In 2017, a nationwide campaign took off over rising fuel prices and the liberalization of PEMEX. The fuel price hike protests (known as the "Gasolinazo") were reportedly the largest nation-wide protests in Mexico in decades.

In sub-Saharan Africa similar patterns emerged. After decades of government price controls and subsidies covering basic consumer goods in the immediate postcolonial era, African states came under structural adjustment pressure to loosen price controls in the 1980s and 1990s. Total foreign debt grew from $61 billion in 1980 to $206 billion by 2002. As Bond (2006: 39) states in stark terms, "By the early 2000s, the debt remained unbearable for at least 21 African countries, at more than 300 percent of export earnings." The foreign debt forced African states to implement structural adjustment policies that broke implicit social contracts with the urban popular classes leading to one of the central grievances of African protest in the late 1980s and 1990s (Bratton and van de Walle 1997). The core coalitions in the African anti-neoliberal protests include public sector unions, school teachers, opposition parties, and university students.

At the continental level, the region witnessed a steep rise in protests since 2000 (see figure 2.2). In the early 2000s, a new wave of popular mobilizations occurred in sub-Saharan Africa over increasing pressures for privatization from a new round of structural adjustment agreements between the IMF, World Bank, and heavily indebted African states. Collective action campaigns took place in Malawi over water privatization led by the state workers' union. In Zambia, multiple mass mobilizations and wildcat strikes were launched against the privatization of state banks and the country's highly valued copper mines (Dwyer and Zeilig 2012). These trends continued through the 2010s. For example, a major protest action over fuel and basic price increases led to university student revolts in several towns in Sudan in both 2012 and 2017, as well as massive protests in Nigeria in 2012. The leading grievances of protests in Africa since 2010 include measures directly related to neoliberal reforms including unemployment, declining access to health and education, and failure of the state infrastructure to deliver social services (Bond 2018).

Across the Middle East and North Africa (MENA) and West Africa, Bush (2010) examined the antecedents to the 2007–2008 wave of food price riots and the key protagonists. In western Africa, major protests occurred in Guinea, Burkina Faso, Mauritania, and Senegal against rising food prices. The protest coalitions included public sector unions, teachers, students, and farmers. They used roadblocks and other disruptive tactics. The demonstrators also incorporated other social citizenship demands of increasing subsidies on basic consumer goods and utilities after decades of adjustment and market deregulation. Similar protest occurred in the MENA countries of Tunisia, Egypt, Jordan, and Yemen (Bush 2010). In Patel and McMichael's (2009: 23) analysis of the global wave of food price hike demonstrations in 2007–2008 in some thirty developing countries, they find the origins of the mobilizations in both structural adjustment and a world food trade regime of neoliberalism as well as the declining commitments of states to the most fundamental material sign of the "social contract"—the provision of the basic food supply.

In late 2019, Iraq, Iran, and Lebanon exploded with civil society protests over the lack of social services and the high cost of living. The protests in Lebanon were the largest outpouring of mass discontent since 2005. In Iraq, protesters faced extreme state repression with over 250 killed and 6,000 injured. In Iran, a fuel price hike led to massive protests in two dozen cities with a reported 115 killed. The Iranian protests erupted in the aftermath of the even larger protests of late 2017 and early 2018 over budget cuts in cash transfers and consumer subsidy reductions that mobilized citizens throughout the national territory in 140 cities. Declining material conditions and growing social inequality galvanized protest campaigns in all three countries as the promises of economic liberalization remain unfulfilled.

In late 2018, a general strike erupted in Burkina Faso led by the National Coalition against Costly Living to battle fuel price hikes. Between December 2018 and February 2019, Sudan experienced over three hundred anti-government demonstrations in the largest wave of protest since independence in 1956 (Morgan 2019). The protests began over sudden fuel price hikes and the subsequent stated repression to quash the unrest has resulted in a reported thirty to fifty deaths. Fuel price hikes in the 2010s have led to major protest episodes in several world regions, from France to Ecuador, Mexico, and Haiti, to multiple states in Africa. In an exhaustive study of individual protest participation across thirty-one African nations between 2002 and 2015, Mueller (2018) finds that those reporting a decline in their living situations in the past year and little confidence in upward mobility were more likely to engage in collective action. She also shows that those embedded in organizational infrastructures such as those residing in cities and those participating in community meetings and religious associations as more likely to participate in protest events.

In South Asia, Uba (2008) has examined dozens of anti-privatization protest events in India between 1991 and 2003. She analyzed anti-privatization mobilizations down to the subnational provincial level and found that the protest campaigns were stronger where public sector unions sided with students and environmentalists in broad coalitions. In a separate study of 108 privatization policy battles in India in the same time period, strike actions by public sector labor unions on average generated two million participants as they fill vital roles in the government's economic infrastructure (Uba 2005). In September of 2016, in what observers call the largest general labor strike in world history, up to 180 million Indian public sector workers rallied against privatization and other neoliberal reforms.

In both the Global North and South, resource infrastructures are one of the most consistent predictors for national-level rebellion against free-market reforms and against the threat of a reduction in material well-being. Resource infrastructures have been defined in both general and specific terms and vary across nation-states. In large cross-national studies, proxy measures are often used to capture the level of resources available for collective action such as urbanization (Walton and Ragin 1990), GDP (Auvinen 1996; Abouharb and Cingranelli 2007), and mineral/natural resource wealth (Wimmer 2013; Arce and Miller 2016). Other studies have focused on more specific resources deriving from civil society, such as labor unions, oppositional political parties, NGOs, and a wide variety of civic associations (e.g., women's groups, student organization, human rights, indigenous groups, etc.).

Spronk and Terhorst (2012) find that social movement coalitions of NGOs, labor unions, and community-based organizations are likely to emerge in campaigns against privatization in the health, electricity, and water/sewage service sectors. Other fungible resources include state infrastructures of hospitals, highways, schools, and social services that were first established in the period of state-led development and not explicitly set up to be appropriated by social movements. As stated above, health, education, state services, and highways all expanded markedly under the developmental state and provide much of the infrastructure to sustain mobilization against economic globalization policies that are perceived as unfavorable. In the neoliberal period, social movements mobilizing against economic liberalization often involve state health workers, civil servants, public school teachers, and students (especially university students), who at times use disruptive protests such as blocking highways until negotiations commence (Almeida 2014, 2015).

Another group of studies focuses on the political context of the countries experiencing economic liberalization and the likelihood of collective action—especially in terms of the level of democracy or political space to organize (Jenkins and Schock 2004). In a multi-decade study, Hendrix and Haggard (2015) show that the economic threat of food price increases is associated with urban protests in Asia and Africa, especially under democratic and semi-authoritarian regimes versus closed and repressive states (see also Berazneva and Lee 2013). Béjar and Moraes (2016) find that protest demonstrations across Latin America are more likely in settings of IMF structural adjustment programs and a low level of political party institutionalization. On a global scale, other studies have shown more collective opposition to economic globalization with deepening democratization at the national level (Arce and Kim 2011) and more agreements with the IMF (Gallo-Cruz 2019). All of the above studies use Polity IV data (Marshall and Jaggers 2009) to construct measures of democratization and regime type. These political context findings with cross-national time series data support scholarship on the growing role of electoral political parties in anti-globalization protests.

The democratic space and low level of institutionalization allow oppositional parties to align with social movements as a means of strengthening their electoral power by mobilizing against economic globalization (Hutter et al. 2018). Political parties act as one of the largest formal organizations available to mobilize citizens in the neoliberal period in which labor unions and rural peasant associations have declined (Almeida 2010a). This trend can be found in several world regions. In Latin America, oppositional political parties have joined in anti-globalization protests, producing large-scale mobilizations in Brazil, Mexico, El Salvador, Honduras, Nicaragua, Costa Rica, Ecuador, Bolivia, Argentina, and Uruguay. These anti-neoliberal mobilizations eventuated in left-leaning electoral victories in nine Latin American countries between the late 1990s and 2010 (Levitsky and Roberts 2011), another consequence of anti-globalization social movements. In Europe, newly created oppositional parties in Spain and Greece have formed symbiotic relationships with citizen movements against austerity policies since the Great Recession (della Porta et al. 2017; Kanellopoulos et al. 2017; Ramiro and Gomez 2017). Indeed, Maria Kousis (2016) found that between 2010 and 2013, left-wing political parties were present in 75 percent of large-scale economic austerity protests in Greece (those with more than five thousand demonstrators).

During this same time period in Portugal, a similar protest wave erupted over austerity and external debt, with the core oppositional infrastructure deriving from public-sector labor unions and leftist oppositional parties in alliance with "new new" social movements of youth mobilizing largely through information and communications technologies (ICTs) (Accornero and Ramos Pinto 2015). A major obstacle facing progressive political parties that are voted into power via anti-neoliberal social movement mobilization is how to maintain the initial energy and emotional effervescence that drove the election campaigns. From Greece to several Latin American left regimes, the conditions that initially led to electoral triumph seem to fade with the passing of time, and these parties in power find it difficult to continue electoral success once they are institutionalized, especially after successive victories.

In the Global North, collective opposition to neoliberalism began with the global economic recession of the early 1970s. By the mid- to late-1970s, major campaigns against economic austerity measures erupted across Europe, including the largest general strikes in France and Italy in decades. In the United States, the protests took the forms of public sector workers (teachers, civil servants, fire and public safety) engaging in short-term protests and strikes as mayors and legislatures slashed city and state budgets. Italy, France, Spain, Sweden Belgium, and Portugal experienced the largest labor strikes in decades in the late 1970s and early 1980s as governments pulled back on social citizenship rights. These popular movement mobilizations in the Global North were most often led by labor unions and reactions to the fiscal crisis of the state (O'Connor 1973).

Between the 1980s and 2000s, neoliberal shifts largely affected the Global North via economic restructuring with the relocation of manufacturing to the semiperiphery and periphery of the global system. This resulted in resistance at the local level to factory closures (Moody 1997) and at the national level to organized opposition to free trade agreements (Kay and Evans 2018). In the twenty-first century, austerity and anti-neoliberal protests in the Global North have growing similarities to earlier waves of such protests in the Global South—especially following the 2008–2009 Great Recession. As figure 2.2 illustrates, this includes the massive austerity protests in the 2010s experienced in Britain, Canada, France, Spain, Belgium, Greece, Italy, and Portugal, as well as the Occupy and living wage campaigns in the United States.

Local Level Responses to Globalization

Another related focus is subnational and local-level variations in social movement responses to economic globalization. This perspective sheds light on how global change processes are interpreted at the local community level and the likelihood that collective mobilization will occur (Auyero 2001, 2006; Almeida 2012). Globalization processes unevenly affect regions within nation-states (Sassen 2008). Auyero (2001: 35) refers to these dynamics as "glocalization," whereby local conditions combine with global forces. In terms of collective action, localities within states will vary in their responses to globally driven influences. Like national-level collective action studies, subnational, and local-level studies also emphasize economic threats, resource infrastructures, and strategic experience. Because of the local or regional level of analysis, subnational studies of globalization and social movements offer fine-grained accounts of how the mobilization processes take place on the ground versus large cross-national studies that tend to aggregate important correlates of rebellion, resulting in more abstracted depictions of key causal dimensions of anti-neoliberal collective action.

In one of the most thorough studies of municipal-level collective action, Trejo (2012) examines 883 indigenous municipalities in Mexico over twenty-six years (1975–2000). He finds that the strongest predictors of indigenous people's protests include trade liberalization/neoliberal policy shifts, local organization networks tied to the Catholic Church, and prior community experience with social movement mobilization. The Zapatista rebellion in southern Mexico that began in 1994 also offers another emblematic case of subnational resistance to globalization (as a response to NAFTA). In an extensive local-level study of the Zapatista rebellion in Chiapas, Mexico, across 111 municipalities over ten years (1994–2003), Inclán (2018) finds that municipalities with past protest experience, more closed governments, and districts with a military presence were more likely to engage in collective action. In two of the largest sustained protest campaigns against privatization in Latin America (in Costa Rica and El Salvador), Almeida (2012) demonstrates that collective protests were more intensive in communities that had public universities, major highways, state administrative offices, NGOs, and left-leaning oppositional political parties. Similar findings have been shown for Guatemala, Nicaragua, and Panama in a larger study of local-level social movement responses to economic globalization in Central America (Almeida 2014).

In Bolivia between 1995 and 2005, Arce and Rice (2009) found that direct foreign investment increased the level of protest at the provincial level, with the battle over control of natural resources serving as a highly contentious issue in indigenous communities. At the height of anti-neoliberal protests in Bolivia in the early 2000s, in a detailed study across the country's 314 municipalities, local-level protests were associated with a larger density of NGOs (Boulding 2014). Whether NGOs are agents of mobilization or demobilization is one of the most polemical debates among scholars who study development, collective action, and globalization. As Subramaniam (2007) contends in her case studies of women's empowerment in rural India, Western NGO donors often control the agendas and priorities for collective action campaigns at the local level. Similar observations have been made about NGOs across Africa (Branch and Mampilly 2015) and South Asia (Bano 2012). Bob (2005) makes a compelling case for the relative success of local NGOs in Nigeria and southern Mexico based on their differing ability to frame community struggles in a manner that is acceptable to sponsors in the Global North. In Krishna's (2002) extensive study of sixty-nine Indian villages in the states of Rajasthan and Madhya Pradesh, he highlights more informal networks of social capital (based on reported community solidarity) as explaining varying levels of local political participation, including protest.

In larger industrializing states such as Argentina, China, Brazil, India, Nigeria, South Africa, and South Korea, subnational opposition to privatization has been led by public-sector labor unions (Sandoval 2007; Uba 2008; Pereyra et al. 2015) and the recently unemployed, and struggles increasingly turn toward more defensive postures of preserving employment and benefits established under state-led development (Rossi 2017). Local-level protests in China have also been driven by economic reforms and associated grievances of job loss, pension arrears, and land access (Lee 2007; Tong and Lei 2013). Community-level mobilizations have largely used the administrative state structure held over from the Mao Zedong era to register petitions (Chen 2012). Systematic studies of rural collective resistance to economic reforms in China in the twenty-first century have found that the most rebellious villages (measured by both petitions and noninstitutional protest) have been characterized by lineal family networks under the economic threat of land annexation, while successful local mobilization is associated with more formal senior associations recognized by the state (Lu and Tao 2017). In the 2010s, South Africa has also witnessed an upsurge in community level protests over delivery and accessibility of basic social services (Paret 2017).

Arce and Mangonnet (2013) examine subnational resistance to economic liberalization in Argentina between 1993 and 2005, focusing on the protest tactic of roadblocks, a strategy that has surpassed labor strikes in contemporary Argentina and is now widely used throughout Latin America and beyond. They show that provinces with strong Peronist oppositional political parties, past collective action, and the threat of high unemployment experienced more roadblocks. Auyero (2007) also demonstrates that collective lootings in Argentina during the 2001 foreign debt crisis were more frequent in localities with strong Peronist clientelist networks.

In the context of the 2008–2009 Great Recession, scholars are examining subnational opposition to the social consequences of the global economic crisis in industrialized democracies in the Global North. For example, Vasi and Suh's (2016) study of the spread of the Occupy Wall Street movement across US cities found that many of the conditions associated with local opposition to neoliberalism in the Global South were also essential in generating movement activity in the North, including the presence of left-leaning parties, universities, and higher levels of past activism. They also found that the presence of a pro-Occupy Facebook page in a city also increased the likelihood of a local Occupy action at the community level. In another subnational study of the Occupy Wall Street movement across Californian cities, Curran et al. (2014) demonstrated that protest encampments were more likely to occur in towns with more votes for Democrats, a large youth population, and universities, and the encampments were negatively associated with the presence of military bases. Nearly a third of Californian towns and cities had Facebook pages for an Occupy encampment in the fall of 2011.

Between 2010 and 2013, during the mass mobilizations against austerity and economic adjustment in Greece, the coordination of simultaneous demonstrations and strikes across geographic space was largely explained by the local presence of labor union chapters, student groups, and leftist oppositional political parties, often in coalition with one another (Diani and Kousis 2014; Kousis 2015, 2016). It is also interesting to note how similar coalitions are molded together by economic threats (McCammon and Van Dyke 2010) in both the Global North and South in sustaining local-level campaigns during periods of globalization-induced crises.

The Case of Local Opposition to Neoliberalism in Costa Rica

One of many cases to systematically examine the dynamics of local level opposition to neoliberalism is the 2018 fiscal reform in Costa Rica. The policy began when an IMF technical team visited Costa Rica in June of 2018 and advised the newly elected government on a series of fiscal reforms—known as the "Combo Fiscal." The government entered the reforms into parliamentary debate in September of 2018 and the legislature approved them in December. The legislation called for a lifting of price controls on consumer products, increasing the sales tax on basic consumer items, and reducing benefits for state sector employees (the kinds of economic liberalization policies that led to major unrest in Chile, Ecuador, and Lebanon in 2019). The Costa Rican state was trying to reduce an ongoing fiscal deficit that had reached 7 percent of GDP. For years, the state had been financing almost half of the national budget via foreign borrowing. Besides steep consumer subsidy cuts and a reduction in state worker benefits, the new fiscal legislation did not increase taxation on large private sector establishments and corporations. The lack of shared burden by the more privileged economic sectors further motivated labor unions and the popular sectors to mobilize a campaign of resistance.

Labor unions and their civil society allies immediately coordinated mobilizations against the neoliberal reforms, forming an ad hoc coalition called "Unidad Sindical y Social Nacional." Every Wednesday in September of 2018 labor and popular sectors organized a massive street march in the capital. On September 26, the anti-neoliberal opposition pulled off the largest street demonstration in Costa Rican history with a reported 500,000 persons in attendance (Cordero Ulate 2019)—approximately 10 percent of the entire national population. Labor unions also initiated a three-month long strike on September 10 and began coordinating protest actions across the national territory in alliance with other groups such as students, community organizations, and other popular sectors (Alvarado Alcázar and Martínez Sánchez 2018; Cordero Ulate 2019).

Figure 2.5 illustrates the distribution and frequency of the protests down to the local level in the first month of the protest campaign (September 10 to 30) over the Combo Fiscal.[6] Two hundred and sixty-six protest events were documented. Figure 2.5 also shows where protests took place over the Central American Free Trade Agreement between 2003 and 2007, a campaign that documented 694 protest events (Almeida 2014). The 2018 fiscal reform protests involved strikes, rallies, street marches and roadblocks. Public sector labor unions served as the vanguard of the campaign. When the strike began to weaken in October of 2018, public school teachers carried the movement forward for another month, refusing to back down (Cordero Ulate 2019).

Table 2.1 provides a multivariate count regression model examining the structural forces influencing anti-neoliberal protest at the local level against the 2018 Fiscal Reform legislation in Costa Rica. Multiple features of the state infrastructure first established in the state-led development period (1940–1980) increased the rate of anti-neoliberal resistance in the

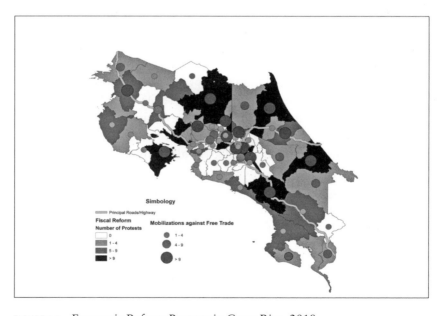

FIGURE 2.5   Economic Reform Protests in Costa Rica, 2018

TABLE 2.1   *Negative Binomial Count Regression Model Predicting the Intensity of Municipal Level Protest against 2018 Neoliberal Fiscal Reforms in Costa Rica*

Independent Variable	Coefficient	Incidence Rate Ratio
State Infrastructure		
Administrative Infrastructure (Provincial Capital)	1.362*** (.276)	3.902
Transportation Infrastructure (Highway in Municipality)	.726*** (.217)	2.066
Education Infrastructure I (Public University)	−.247 (.283)	.781
Education Infrastructure II (Public Schools)	.011*** (.002)	1.011
Strategic Experience/Capital Protests against CAFTA (2003–2007)	.007*** (.002)	1.007
Population Size (ln)	−.005 (.184)	.995
Intercept	−.386 (1.941)	
Log likelihood	−144.778	
Pseudo R²	.17	
N	81	

Note: Robust Standard Errors are in parentheses.
*** $p \leq .001$ (two-tailed tests)

present. Provincial capital cities act as a proxy for administrative infrastructure. More protest events took place where government offices are housed as a target for demands. Municipalities (or *cantones*) that are transected by one of the country's two main highway systems also experience increased rates of protest. This highway infrastructure was also built in the period of welfare state expansion. As is common throughout Latin America and the greater Global South, subaltern groups use the roadblock on major thoroughfares to apply pressure on elite economic and political adversaries to negotiate by disrupting traffic and commerce.[7]

The more public schools in a municipality also intensified collective action against the neoliberal reforms. As mentioned above, school teachers played a vital role in resisting the economic liberalization measures and were geographically concentrated near their work locations. The four prominent teacher unions were unified throughout the campaign. Strategic experience or past collective action is also a powerful force to initiate new campaigns and rounds of opposition to free market forms of globalization. Between 2003 and 2007, Costa Rican labor unions, ecologists, students, and teachers launched the largest campaign in the region against the Central American Free Trade Agreement (CAFTA). Even more than ten years after the height of the CAFTA campaign, the movement had still deposited enduring mobilization knowhow in local communities. Those localities that mobilized against CAFTA in the 2000s, participated in the resistance against the 2018 fiscal reforms with significantly higher intensity than those that mobilized less.

The case of Costa Rica shows how local level resistance materializes against neoliberal reforms. Those structures established in the previous period of state-led development—such as mass public schooling, highways, hospitals, and administrative offices—provide the capacity and motivations to launch campaigns in the present when social citizenship rights come under threat or erosion. Those communities with past strategic experience in coordinating resistance may be able to resurrect a protest campaign much easier than localities that lack such experience. At the same time, it is critical to acknowledge that there exists critical variations and histories of state-led development in different world regions and nations whereby particular state structures and oppositional actors may be more prominent than others. At the same time, the case of Costa Rica reminds us of the relevance of the state sector as one of the few viable structures stretching across the national territory capable of providing a fungible organizational infrastructure to oppositional groups. Currently, several mega-infrastructural projects are planned throughout the Global South, including China's Belt and Road Initiative, the African Development Bank's infrastructure development program, and the Initiative for the Integration of the Regional Infrastructure of South America (Schindler and Kanai 2019). Such initiatives will likely change the mobilization potential for collective action in these regions.

Types of Neoliberal Policies

Given the variety of neoliberal measures, scholars currently debate which specific economic liberalization policies have been more likely to result in protests and rebellions at both the local and national levels. Baker (2009), using extensive Latin American public opinion data, finds that privatization is much less popular than free trade policies. Lindh (2015) also finds generally unfavorable attitudes toward the privatization of social services across OECD countries. Some of the largest anti-neoliberal protest demonstrations and general strikes across eastern Europe, Asia, and Latin America have centered on the privatization of the basic social and economic infrastructure, including public health systems, social security, water administration, and electrical power distribution. As recently as 2019, a massive uprising against health care and education privatization took place in Honduras (Sosa and Almeida 2019).

In Africa, the global turn to market deregulation has resulted in the largest mobilizations centering on the costs of food, transportation, fuel, and the general loss of social services (Sneyd et al. 2013; Branch and Mampilly 2015). In early 2012, Nigeria experienced one of the largest mass mobilizations in decades over an IMF-advised policy cutting fuel subsidies. Occurring at the same time of the US protests, the movement was coined Occupy Nigeria (Branch and Mampilly 2015). Mexico experienced similar nation-wide protests over fuel subsidy cuts in early 2017 as did Ecuador in 2019 (mentioned above).

Free trade policies and treaties have proliferated since the establishment of NAFTA in 1994 and the WTO in 1995 (Dicken 2015) and have consequently sparked massive protest campaigns in Argentina, Brazil, Germany, Ecuador, El Salvador, Canada, Costa Rica, Colombia, France, Guatemala, Italy, Peru, United States, and South Korea; they also served as the initial impetus to the 1994 Zapatista rebellion in southern Mexico. Moreover, one of the most successful transnational protest campaigns in contemporary Latin America involved the effective coordination of labor unions and leftist political parties across South America to defeat the Free Trade Area of the Americas (FTAA) (von Bülow 2010; Herkenrath 2011; Silva 2013). Other cross-national comparative work has demonstrated that the timing of economic liberalization matters for mobilization potential and movement strategy. Civil society is more likely to coordinate protest campaigns after several rounds of negative policy experiences with neoliberalism, and activists and civic leaders develop more efficacious mobilizing strategies in the wake of past oppositional defeats (Almeida 2014). Recent studies of collective resistance to market-driven globalization have also moved beyond resource infrastructure and political context explanations to focus on the moral meanings of economic reforms for local populations using ethnographic research strategies (Auyero 2006; Simmons 2016).

This chapter brings focused attention on the basis of anti-neoliberal opposition and how that opposition varies across nation-states and localities within countries. A historically grounded interpretation of contemporary events highlights both the motivations and the capacity to sustain collective action campaigns against neoliberal forms of economic development. The prior period of state-led development is not simply a past form of economic strategy. Rather, the era of state-led development fostered an unprecedented expansion of the core infrastructure of nation-states in the developing world accompanied by a fledgling welfare state. Between the 1930s and the 1980s, the project of the developmentalist state brought millions of people around the globe into mass education institutions and provided access to basic social services and health care. The redistributional component of the developmentalist state extended social citizenship rights. At the same time, it raised the mobilization potential of civil society to launch collective action campaigns to extend social and political rights by creating a fungible infrastructure to coordinate and sustain social movement–type activity.

With the transition to neoliberalism and the debt crisis of the 1980s, the social citizenship gains of state-led development increasingly came under threat via structural adjustment, austerity, labor flexibility, privatization, and the loosening of price controls. While even under state-led development, women, ethnic minorities, and indigenous people received less protections, these social citizenship threats provided the catalyst for struggles to protect civil society from the new harms associated with the deterioration in social welfare. The actual physical and organizational structures persisting from state-led development into the neoliberal period have provided an array of resources to sustain oppositional campaigns to orthodox market reforms.

From this broad framework, a more refined research agenda can be undertaken to more precisely understand the pace of neoliberalization and its alternatives by examining the infrastructural histories of nation-states and local regions. Analogous to path dependent studies on how initial economic structures become institutionalized and shape future development trajectories (Mahoney 2010), different types of state-led development schemes in the mid-twentieth century will likely condition the forms of resistance to neoliberalism in the twenty-first century as well as related struggles to turn back planetary warming.

In terms of scale, we would likely observe greater levels of neoliberal opposition when larger welfare states and state-directed economies begin to reverse social citizenship rights. This same proposition should also be investigated across localities *within* nation-states as the state infrastructure is unevenly deposited across territorial districts as we have shown with the case of Costa Rica. The types of state development political regimes will also likely condition the social actors and sectors leading the opposition to neoliberalism decades later.

The institutional outcomes of these anti-neoliberal campaigns also vary. Many end in defeat, as witnessed by the ongoing intensification of economic liberalization in the twenty-first century. Nonetheless, at times, campaigns turn back the original neoliberal policies such as an IMF measure, a price hike, or an attempt at privatization. In more proactive struggles, leaders of anti-neoliberal social movement campaigns have converted the struggle into electoral triumphs at the local and national level via progressive political parties. Harris and Scully (2015) document another related anti-neoliberal trend or Polanyi-type "counter-movement" pushed from below over the past two decades in the rapid growth of social assistance programs in the Global South. States have enrolled up to one billion citizen beneficiaries in these initiatives. Such programs enacted by progressive and pro-neoliberal states demonstrate how elites adopt policies in response to rebellions from below, even in indirect terms. We continue to show these interactive processes between movements and global institutional change in the following chapters.

THREE

Transnational Movements

Climate Justice

···

As described in chapter 2, the transition to the neoliberal form of global capitalism in the late twentieth century corresponded with a variety of novel forms of resistance at the local, national, and international levels of political life. Neoliberalism produces new models of unequal development between the capitalist core and periphery (Amin 1976; Bond 2006) as well as within nation-states along with a host of tensions and threats motivating popular movements. These struggles will likely intensify as we move into the third decade of the new millennium. At the local level, collective action centers on everyday forms of resistance and grassroots struggles over land grabbing, mining, and mega development projects (Almeida 2019b). At the national level, opposition to neoliberalism manifests in the form of social movement campaigns against a bundle of economic liberalization policies that include austerity cuts, consumer price hikes, free trade agreements, privatization, deregulation, and labor flexibility laws (Walton and Seddon 1994; Silva 2009). At the transnational level, mobilization against global capital is most pronounced in the Economic Justice movement, the World Social Forums, and, increasingly, the movement for climate justice, the focus of this chapter.

In past decades, sociologists theorized that global capitalist accumulation would create its own self-induced limits through the depletion of natural resources, pollution, and environmental destruction (Schnaiberg 1980; Gould, Pellow, and Schnaiberg 2004). James O'Connor (1988) conceptualized these processes as the "second contradiction of capitalism," a contradiction in addition to the capitalist crisis of overproduction. In this perspective, advanced forms of capitalist accumulation undermine the necessary material requisites for system reproduction by destroying the ecological bases ("the condition of production") for continuous and expanded industrial activities on a global scale, leading to a crisis of underproduction. More recently, scholars contributing to these debates incorporate carbon emissions and global warming as a core "ecological rift" associated with global capitalism (Foster, Clark, and York 2011; Moore 2015).

The most recent scientific reporting suggests that the outlook for continued global warming is dire. Instead of a reduction in carbon emissions over the past two years (2017–2018), there has been a global increase by 1.6 and 2.7 percent, respectively (Dennis and Mooney 2018). Moreover, the past four years (2015–2018) represent the warmest mean global temperatures on record, and July, September, and October 2019 marked the hottest months ever registered.[1] The twenty warmest years on record have occurred over the past twenty-two years (World Meteorological Organization 2018). This particular environmental challenge of global warming and climate change, upheld by a variety of neoliberal forms of capitalism in the twenty-first century, is also generating one of the most extensive transnational movement in history—the movement for climate justice. Environmental justice and climate justice combine threats of environmental degradation with concerns about inequality and the larger impacts on disadvantaged populations (Bullard 2005; Harlan, Pellow, Roberts, and Bell 2015; Pellow 2017).

Institutional Origins of the Early Climate Justice Movement

In the previous chapter, we focused on *economic threats* and the erosion of social citizenship rights. *Ecological threats*, such as climate change, provide another major incentive for collective action in that failure to mobilize in the present will likely lead to worsening environmental conditions (Johnson and Frickel 2011; Almeida 2018). Earlier conservation movements (often involving more privileged social strata) organized in waves of environmentalism since the late nineteenth century against ecological threats associated with the expansion of industrial capital (Gottlieb 1993). The global movement against environmental threats and ecological destruction has been active since at least the first Earth Day in 1970 with coordinated actions by millions of people across the planet. This annual event also reached a massive scale of citizen participation in 1990, 2000, and 2010, respectively (Ciplet, Roberts, and Khan 2015).

After several decades of the scientific community making advances in empirically establishing the growth of CO_2 emissions and global warming between the 1950s and early 1980s, the movement to resist the ecological threat of climate change emerged. Beginning with the World Meteorological Organization's first World Climate Conference in 1979, climate scientists and environmental NGOs started to push global bodies and nation-states to take action based on atmospheric studies demonstrating a clear trend in global warming and its negative ecological consequences. At this time, the United Nations founded the Intergovernmental Panel on Climate Change (IPCC) to begin global discussions to reduce greenhouse gas emissions. Concurrently, a global network of environmental NGOs emerged to pressure the UN for a binding international climate accord—the Climate Action Network (CAN) (Brecher 2015). During the United Nations Conference on Environment and Development (Earth Summit) in Rio de Janeiro, Brazil, in 1992, the United Nations Framework Convention on Climate Change (UNFCC) was established as an intergovernmental forum to work on reducing global warming (Caniglia et al. 2015). In 1995, the UNFCC forum also set up annual meetings with UN member countries to further make progress on a global climate treaty to decrease carbon emissions—the Conference of the Parties (COP). Throughout the 1990s and early 2000s, the Global Climate movement to reduce greenhouse gases was concentrated in advanced capitalist countries and largely worked through the institutional channels of these UN bodies via the participation of environmental NGOs. This period has been referred to as "mobilization from above" (Brecher 2015).

Starting in the mid-2000s, the global Climate Justice movement became more contentious organizing rallies and marches across the globe, usually in concert with major United Nations Climate Summits and COP meetings (Garrelts and Dietz 2014). The use of more noninstitutionalized tactics rose in tandem with the lack of progress within the UN system to enforce past agreements and hold countries accountable for CO_2 emissions. Already, by 2005, the movement was capable of mobilizing simultaneous demonstrations in cities across several continents (see table 3.1). The Climate Justice movement first peaked between 2009 and 2015 by holding global days of protest in a majority of countries and sustained another campaign in September of 2018. This global reach marked the transnational Climate Justice movement as the most extensive social movement on the planet in terms of the capacity to hold multiple and simultaneous global actions.

In 2019, new initiatives with global ambitions infused new life into the Climate Justice movement, namely, Extinction Rebellion and Fridays for Future. Extinction Rebellion was founded in the United Kingdom by a small group of activists in late 2018. The movement focuses on the use of disruptive and dramatic tactics to bring greater attention to the issue of planetary warming. The movement has multiplied rapidly with affinity groups organized on every continent. One prominent activist summarized the rapid diffusion of Extinction Rebellion: "In April 2019 we began our first phase of International Rebellion. In Pakistan, we marched through the capital. In the US, we glued ourselves to a bank. In the Netherlands, we occupied The Hague. In Austria, we blocked roads. In Chile, we lay down in the middle of a street. In Ghana, we blew whistles to sound the climate alarm" (Extinction Rebellion 2019: 10).

An even more extensive push to expand the Climate Justice movement globally is found in the youth climate strikes first promoted by Swedish teenager Greta Thunberg (Thunberg 2019). The climate strikes began in 2019 with school walkouts across Europe and then later to the rest of the world. From primary schools to high schools, youth have walked out of school every Friday to pressure their respective governments to act more assertively on reaching a global climate policy. The movement is called Fridays for Future. Larger Global Days of Action have also been organized by Fridays for Future in 2019 surpassing the levels of climate justice mobilization in 2014 and 2015 (de Moor et al. 2020). Indeed, in September 2019, Fridays for Future (and partnering organizations) held a week of street actions and peaceful demonstrations across the planet reaching a reported 185 countries with over six thousand events and 7.6 million participants. As part of this campaign, Greta Thunberg led a street march in New York city with 250,000 demonstrators (Almeida 2020). Activists timed the worldwide campaign to place pressure on the United Nations Climate Action Summit occurring in New York in late September.

The focus here is on the organizational infrastructure that has made the transnational climate movement so extensive and prospects for future mobilization and lasting coordination among popular organizations and movements. In particular, we examine the role of the Global Economic Justice movement and the Anti-War movement in providing the initial infrastructural and experiential basis for planetary mobilization against climate change.

Global Justice and Anti-War Movements

The Global Economic Justice movement (or Alter-Globalization movement) took off in the late 1990s shortly after the establishment of the World Trade Organization (WTO) in 1996. The movement quickly developed an innovative organizational template of mobilizing on a transnational level. The coordinating template involved mobilizing a massive series of actions at the focal conference/summit/financial meeting while simultaneously holding dozens of solidarity actions across the globe (Almeida and Lichbach 2003). This particular transnational organizing model is referred to by activists as a "Global Day of Action" (Wood 2004). The movement was a response to the neoliberal form of global capitalism that had been taking shape since the 1980s with a heavy emphasis on free trade (Dreiling and Darves 2016) and loosening of social and environmental protections at the national level (i.e., the erosion of social citizenship discussed in chapter 2). The emerging Transnational Economic Justice movement began to take advantage of the rise of internet communication technologies (ICTs) increasingly available in the late 1990s—especially web pages, email listservs, and cellular phones. Beginning with international financial meetings in Europe in the late 1990s, G8 summits in Birmingham and Cologne (Wood 2004), and the 1999 WTO conference in Seattle (Smith 2001), the global organizational template became institutionalized in world society. Indeed, by the turn of the twenty-first century, the Global Economic Justice movement organized over fifteen transnational campaigns per year with over 200,000 participants (Lichbach 2003; Wood 2004; Juris 2008).

By the early 2000s, the Global Economic Justice movement expanded the simultaneous protests to every continent. This would become the main form of transnational opposition to global capitalism in the twenty-first century (Wood 2012). After the WTO meetings in Seattle, global days of protest took place between 2000 and 2003. These included the IMF–World Bank meetings in Prague in September 2000, the G8 conference in Genoa in 2001, the WTO ministerial in Doha, Qatar in November 2001, and the fifth WTO Ministerial in 2003 in Cancun, Mexico (Juris 2008). The Global Economic Justice movement brought a wide coalition of groups into their Global Days of Action campaigns such as youth, leftist and green political party militants, labor unions, environmentalists, LGBTQ groups, indigenous peoples, feminists, anarchists, among many others. They united around the idea of protecting social citizenship and environmental rights granted by nation-states in the twentieth century and now under threat with an emergent neoliberal form of global trade and economic integration.

Figure 3.1 demonstrates the global reach of the protests against the IMF–World Bank meetings on September 26, 2000. Employing the Global Day of Action template (in this case called "S26"), the Alter-Globalization movement organized major demonstrations at the core meeting in Prague, Czech Republic, as well as in at least forty-three countries with events on every continent.[2] Hence, already by turn of the new millennium, activists effectively erected a global infrastructure that would be used to sustain transnational mobilizations for the next two decades.

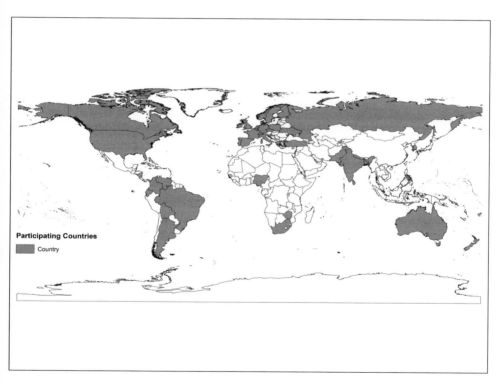

FIGURE 3.1   Global Protests against IMF–World Bank Meetings in Prague, Czech Republic, 2000

The Global Economic Justice movement spilled over into the Global Anti-War movement in the mid-2000s (Fisher 2007). The same template was used by the movement to try and prevent and end the war in Iraq. For example, one of the largest protests in world history took place on February 15, 2003, against the impending US invasion of Iraq. Nearly eight hundred cities in eighty countries (involving some ten to fifteen million people) participated in demonstrations against initiating a war on Iraq using the Global Day of Action coordinating structure (figure 3.2). The issues and networks involved in the Global Justice movement continued via the World Social Forum and mass demonstrations outside G20 meetings in the 2010s, as well as the Global Day of Action in October of 2011 at the height of the Occupy Wall Street campaign.

Indeed, many of the leaders and groups organizing the February 15, 2003, Global Day of Action against the imminent US invasion of Iraq

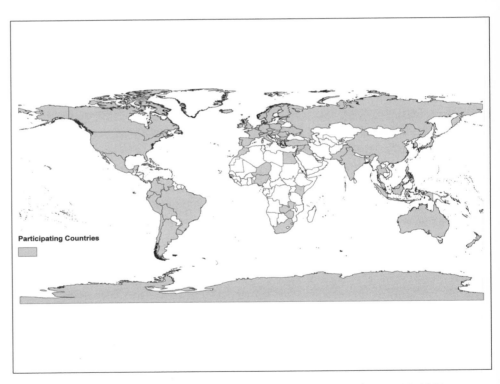

Participating Countries

FIGURE 3.2   Countries with an Anti-War Demonstration on February 15, 2003

came out of the Global Economic Justice movement (Wallerstein 2003a). While past examples exist of a Global Day of Action such as May Day (International Workers' Day), International Women's Day, and Earth Day, the early 2000s marked a major shift in world history in terms of the scale and level of contention in internationally coordinated collective action. ICT technologies developed between 1999 and 2003, made it possible to plan simultaneous events by establishing multilingual web sites and email listservs to recruit organizations and sympathetic populations to participate in synchronized transnational protest events across the planet (Almeida and Lichbach 2003).

Transnational Organizational Infrastructure and Strategic Experience

Just as social movement scholars have demonstrated the role of preexisting organizations and institutions in sustaining social movement emergence at the community and national levels of political life (Morris 1984; McAdam 2003; Andrews 2004), transnational structures serve a similar function in supporting campaigns at the international level in the twenty-first century. The transnational organizational infrastructure is loosely coupled and supported by an ever-changing conglomeration of ad hoc coalitions, groups, and associations such as the Association for the Taxation of Financial Transactions and for Citizens Action (ATTAC), Greenpeace, Friends of the Earth, Peoples Global Action, the European Social Forum, the World Social Forum, Avaaz, 350.org, international labor unions, and many important others.

In addition, the strategic capacity (Ganz 2009) and collective action experience of the actors within the transnational organizational infrastructure allow for future rounds of internationally coordinated events. This is because solidarity relations have been established across international borders and intercontinental collective action has succeeded on dozens of occasions (Smith 2008). The transnational activists in the Alter-Globalization and Anti-War movements have learned to reprogram ICTs more in their favor as a form of mass self-communication to influence the minds of global civil society in a struggle with traditional and corporate-controlled media sources (Castells 2013).

The Contemporary Climate Justice Movement

Similar networks of transnational activists began to piece together the first Global Days of Action to reduce carbon emissions in 2005 and 2006. In 2005, climate activists called for a day of global protests on December 3 (known as "D3") to coincide with the eleventh Conference of the Parties (COP 11) UN climate summit in Montreal. In 2006, the international movement launched another day of global action (called N4) on November 4 in correspondence with the COP 12 climate conference in Nairobi, Kenya. These global networks came out of the Alter-Globalization and Anti-War movements of the early 2000s to now battle climate change (Bond 2012) and used the same symbols for mobilization appeals such as the precise date of the actions as the rally cry ("N4," "D3," etc.). The incipient Climate Justice movement used the same template for organizing as the Global Economic Justice and Anti-War movements—focus on large demonstrations at the main event where the conference is taking place combined with dozens of solidarity events around the world (Pleyers 2010).

Between 2005 and 2008, the Global Justice and Anti-War movements were joined by coalitions such as the Campaign against Climate Change and transnational environmental NGOs such as Friends of the Earth and Greenpeace (Foran 2014). By 2009, the Climate Justice movement reached 143 nations in the days of global action leading up to COP 15 in Copenhagen. The global reach was carried out with the assistance of more assertive coalitions such as Climate Justice Action and Climate Justice Now! with greater representation from the Global South (table 3.1). Hence, a major leap in scale in the history of climate justice mobilization occurred in 2009. Transnational environmental organizations combined a series of actions into a campaign between September and December to produce an unprecedented level of global participation for climate justice. These events included "A Global Wakeup Call" in September, an "International Day of Climate Action in October," and "Real Deal Vigils" in December to pressure negotiators in Copenhagen.

In the 2010s, environmental NGOs with a strong presence and ground game such as 350.org and Avaaz took a leadership role as brokers in coordinating the large mobilizations. These new transnational brokers captured and sustained the energy of the break though climate mobilizations of 2009 by coordinating major Global Days of Action in 2010, 2011, and 2012. The events were not purely protests but also a variety of proactive collective actions. For example, in October 2010, 350.org coordinated "A Global Work Party." The "work party" included creative collective actions around the world from a bike clinic in Calgary, Alberta, Canada, to Earth prayers in Shrangi-La, China, and a youth climate summit in Addis Ababa, Ethiopia. In total, 142 countries documented participating in this global day of climate action. Equally impressive global days of climate action were held in September of 2011 and May of 2012. Instead of a massive Global Day of Action in 2013, global climate activists used the year for building the movement through a series of workshops and conferences, emphasizing youth participation through broad international coalitions such as the Global Call for Climate Action and Global Power Shift. The on-the-ground organizing paid off by the massive outpouring of climate activism that took place in the Global Days of Action in 2014 and 2015 leading up to the Paris Climate Agreement (table 3.1).

The 2014 and 2015 global days of climate action reached up to 75 percent of all countries on the planet with at least 1.5 million participants

TABLE 3.1   *Number of Countries Participating in Global Days of Climate Justice Action, 2005–2019*

Year	Europe/ North America/ Australia and New Zealand	Asia/ Pacific	Africa	Latin America/ Caribbean	Total Countries
2005	20 (57%)	6 (17%)	3 (9%)	6 (17%)	35
2006	23 (64%)	5 (14%)	4 (11%)	4 (11%)	36
2007	32 (44%)	19 (26%)	13 (18%)	9 (12%)	73
2008	33 (48%)	15 (22%)	11 (16%)	10 (14%)	69
2009	40 (28%)	50 (35%)	25 (17%)	28 (20%)	143
2010	44 (31%)	46 (32%)	28 (20%)	24 (17%)	142
2011	32 (35%)	33 (36%)	13 (14%)	14 (15%)	92
2012	27 (28%)	38 (40%)	19 (20%)	12 (12%)	96
2014	41 (27%)	46 (31%)	35 (23%)	29 (19%)	151
2015	43 (29%)	45 (30%)	37 (25%)	23 (16%)	148
2018	28 (31%)	30 (33%)	20 (22%)	13 (14%)	91
2019	42 (42%)	25 (25%)	13 (13%)	19 (19%)	99

Sources: 350.org; Avaaz; People's Climate Campaign; and FFF. Data for 2019 is only through the first six months of the year.

in each campaign. The events in September 2014 coincided with the United Nations Climate Summit in New York City. Once again, the Climate Action movement maximized the economic justice Global Day of Action template with an extraordinary mass march in New York City with a reported 400,000 participants—perhaps the largest street demonstration in the history of the city. The global solidarity protests also stunned observers by mobilizing nearly two thousand events in 153 countries. The mobilization in November and December of 2015 reached nearly the same levels of international participation leading to the Paris Climate Agreement (see the photo from a local action in the 2015 campaign from Costa Rica in figure 3.4). Another push to make the Paris Agreement binding took place in September of 2018 in a global day of climate action. Finally, as mentioned above, Extinction Rebellion and Fridays for Future introduced new life and energy into the transnational movement in 2019 with a series of even broader globally coordinated actions with high levels of youth participation.

Table 3.1 demonstrates the increasing participation from countries across the globe in the transnational climate justice actions, including from the Global South. This increasing participation by the Global South has occurred in the context of declining costs of internet and cell phone access for the urban working classes in the developing world in the 2010s. Indeed, World Bank data show that only about 30 percent of the population had access to the internet in the advanced industrialized capitalist countries in the year 2000, while less than 10 percent of the rest of the world had access. By 2010, about 28 percent of the world's population enjoyed regular internet access, while jumping to 46 percent by 2016.[3] In addition, over 20 percent of transnational social movement organizations in the Global South were working on environmental issues between 2000 and 2013 (Smith et al. 2018). Such an unprecedented and loosely connected global infrastructure provides a base for future rounds of progressive collective action.

Table 3.2 shows the relationship between the early Alter-Globalization and Anti-War movements and the Transnational Climate Justice movements using a multivariate count regression model with a sample of 181 countries. It examines eleven major transnational campaigns for climate justice that occurred between 2005 and 2018. The dependent variable is the number (or count) of transnational climate justice protest

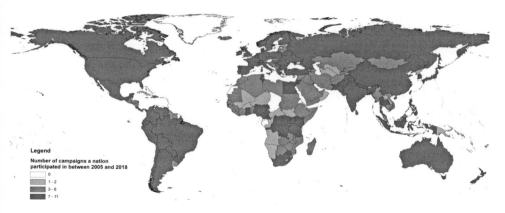

Legend

Number of campaigns a nation
participated in between 2005 and 2018

- 0
- 1 - 2
- 3 - 6
- 7 - 11

FIGURE 3.3   Nations with Citizen Participation in Global Campaigns for Climate Justice, 2005–2018

campaigns in which a country reported participating in with a collective event (ranging from zero to eleven campaigns).[4] Figure 3.3 illustrates the distribution of global participation in the campaigns listed in table 3.1 and analyzed in the count regression model in table 3.2.

Table 3.2 provides support that the internet infrastructure is a transnational resource for the Climate Justice movement to appropriate and mobilize collective action. The more connected individuals are to ICTs in a country, the more the country participates in transnational climate justice campaigns with reported protest events by civil society. The visible ecological threat of sea level rises and catastrophic events such as flooding and hurricanes in low-lying regions (Roberts and Parks 2006) also generated greater levels of climate justice participation. The less visible ecological threat of CO_2 emissions was not significant. In addition, both the Global Economic Justice and the International Anti-War movements experience of organizing transnational collective action in the past, increases the rate of participation in the Climate Justice movement by 35 percent and 29 percent, respectively (holding other variables constant). Hence, the Global Day of Action template becomes deposited in world society as an organizational framework to mobilize future rounds of collective action to demand immediate solutions to the ecological crisis of global warming and climate change.

Countries with larger populations also participated with greater frequency in campaigns to reduce Greenhouse gas emissions. In such cases,

TABLE 3.2 *Poisson Count Regression Model Predicting Protest Participation by Country in Global Climate Justice Campaigns, 2005–2018*

Independent Variable	Coefficient	Incidence Rate Ratio
Internet Infrastructure		
Internet Users per 100 (2005)	.005***	1.005
	(.001)	
Visible Ecological Threats		
% of Population Living Under	.004*	1.004
Five-Meter Elevation (2000)	(.002)	
Non-visible Ecological Threats		
CO_2 Emissions (2005)	−.005	.995
(metric tons per capita)	(.003)	
Transnational Strategic Experience		
Global Economic Justice Protest (2000)	.296***	1.345
Anti-War Protest (2003)	(.063)	
	.253**	1.288
	(.080)	
Population Size		
Population Size (ln) (2005)	.131***	1.140
	(.017)	
Intercept	−.716**	
	(.274)	
Log likelihood	−386.595	
Pseudo R^2	.21	
N	181	
	(Nations)	

Note: Robust Standard errors are in parentheses.
* $p < .10$
** $p < .01$
*** $p < .001$ (two-tailed tests)

the larger the population, the greater the sympathy pool to mobilize. In short, citizen participation in global days of climate action is much stronger in nations with greater access to ICTs, previous experience in global economic justice and anti-war activism, under greater visible ecological threat, and with a relatively larger population size.

The next steps to solidify this infrastructure would be to continue to coordinate global summits and forums with representatives from the participating groups in the Global Days of Action. Past examples include the World People's Summit on Climate Change and the Rights of Mother

FIGURE 3.4 Local Action in San Pedro, Costa Rica, on November 29, 2015, during Global Climate March. *Photo by Paul Almeida.*

Earth (Pachamama) held in Bolivia in 2010 following the worldwide mobilizations associated with COP 15 and the WSF. The Bolivia Summit called on ecological reparations for the global periphery and an immediate and drastic reduction in carbon emissions (Smith 2014).

Perhaps most pressing would be to increase the rate of summits and forums that bring together representatives from the climate justice coalition. The impressive scale of the transnational mobilizations over the past ten years is still limited by the vast amount of time between the launching of Global Days of Action campaigns, even though much traditional organizing takes place on the ground in the interim periods. To overcome the "flash activism" nature of these campaigns and to build the necessary level of solidarity between diverse groups, classes, and sectors for a long-term Anti-Systemic movement (Amin 1990; Ciplet, Roberts, and Khan 2015), climate justice activists need to continue to find avenues and mechanisms for more frequent forums and mobilizations that can maintain and accelerate the momentum of a truly planetary movement. This appears to be the trend with the rise of more sustained and assertive climate justice mobilization in the 2020s.

The increasing intensity of climate change as an existential ecological threat does create *relatively* more favorable conditions for international unity and avoid the sectarianism and fragmentation observed in previous attempts at building an *internationale* or permanent global organization of progressive sectors and groups (Amin 2019). The environmental threat is imminent and global, providing urgency and aligning common interests, the basic building blocks of sustained collective action (Almeida 2019b). At the same time, a number of preexisting social and economic divisions will need to be given heightened recognition to build enduring transnational coalitions across the lines of race, class, gender, and colonial status. The environmental justice movement against ecological racism (Bullard 2005; Roberts, Pellow, and Mohai 2018), the Cochabamba Climate Change conference (Bond 2012), and the current mass mobilizations fostering intersectional alliances (Luna 2016; Terriquez et al. 2018) offer some of the best models to incorporate within the larger global Climate Justice movement (Bhavnani et al. 2019). The movement appears to be moving in that direction by building coalitions with Black Lives Matter and gender-based groups in the United States and major labor confederations and indigenous groups in the Global South.[5]

With global warming disproportionately harming billions of the world's poor and excluded by global capital, the Climate Justice movement should continue it's extraordinary efforts to organize the most vulnerable groups and strata in the Global North or South. Chase-Dunn and Reese (2007) demonstrate that previous progressive parties organized on a global scale were initially able to coordinate simultaneously in the global periphery and capitalist core with membership from a variety of social sectors, including peasants and the urban working class. The transnational climate justice alliance may also build internal cohesion by mobilizing against the xenophobia, authoritarianism, and climate change deniability of right-wing populism. The stakes could not be higher for the Climate Justice movement to achieve immediate victories in reducing greenhouse gas emissions and avoid a planetary ecological collapse.

FOUR

The New Global Left and the World Revolution of 20xx

Movements, Culture, Fronts, and Organization

....

A brief overview of the history of world revolutions in the Europe-centered world-system since the Protestant Reformation provides the context for comprehending the nature of the New Global Left that has emerged since 1994. Our analysis of the nature of the contemporary network of progressive transnational social movements is based on studies of political ideas contained in documents produced by social movement organizations participating in the World Social Forum (WSF) process and surveys that were done of attendees at meetings of the WSF. We also report the conclusions of William Carroll's (2015a, 2016) studies of transnational alternative policy groups. We provide an overview of the Social Forum process as an important venue for studying the New Global Left. We also further discuss the nature of contemporary global civil society and the place of the New Global Left within it. We consider the potential for a network of progressive social movements to come together to play an important role in world politics in the coming decades of the twenty-first century. The histories of united and popular fronts are particularly relevant to contemporary and near-future situations. The world-systems perspective sees the evolution of global governance and the capitalist world economy as driven by a sequence of world revolutions in which local rebellions that are clustered together in time pose threats to the structures of global power.

We consider the potential for transnational social movements and progressive regimes to transform the capitalist world-system into a more humane and democratic world society within the next fifty years. To investigate this potential, we focus on the interconnections between existing

---

The contemporary world revolution has been called "20xx" by Chase-Dunn and Niemeyer (2009) because they were not sure what year would be best for signifying the meaning of the assemblage of rebellions and movements that have emerged since the last decade of the twentieth century.

movements and the processes by which movements have merged, collaborated, and articulated in the past. We also discuss the potential for the formation of capable organizational instruments that can have consequences for the nature of the world-system in the next decades.[1] The general logic of coalition formation is considered and the literature on coalitions within and among social movements is reviewed. The focus here is on the whole world polity but also the much larger literature on movement coalitions within national societies is considered with an eye to implications for understanding processes of convergence and divergence among transnational social movements (Dale 2011). The histories of united fronts and popular fronts are considered as to their relevance to the contemporary and near-future situation. The contentious relationship between anti-systemic social movements and reformist governments is also considered. And we hazard some guesses about which of the existing social movement coalitions might be able to forge a more formidable assault that could seriously alter the existing institutions and structures of the world-system.

We focus on transnational social movements (movements organized in more than one country) in the context of global civil society to investigate the potential for a network of radical social movements to come together to play an important role in world politics in the next few decades. Important institutional changes in the modern world-system have been spurred by social movements in the past. The New Deal was given a powerful shove by the labor movement, including socialists, communists, and anarchists, in the 1930s. The problem addressed here is how a powerful coalition of anti-systemic movements might once again become an important force in the context of the crises that are emergent in the twenty-first century by forging a global organizational instrument that would coordinate local, national, and transnational anti-systemic movements.[2]

The Europe-centered, modern world-system has been a multicentric hierarchical structure of competing firms and states in which capitalism became the predominant logic of accumulation in the long sixteenth century (Wallerstein 1974, 2004). Global governance was organized by a series of hegemonic core powers, the Dutch in the seventeenth century, the British in the nineteenth century, and the United States in the twentieth century. The expansion and deepening of this system were driven by a series of world revolutions in which local rebellions that were clustered together in the same decades posed powerful threats to the rule of the "great powers" and predominant global elites (Wallerstein 2004). Contenders among those global elites who wanted to maintain their privileges and power, or those that sought such heights, had to figure out how to manage world order in the context of powerful rebellions from below (Beck 2011). The Protestant Reformation was such a world revolution in the sixteenth century and it played an important part in the rise of Dutch hegemony. The world revolution of 1789 included the French Revolution, but also the successful independence struggles in the United States and Haiti, and then in most of the colonies in Latin America (Polasky 2016). World revolutions are named by a symbolic year in which some of the major events that indicate the nature of the revolts occurred.

The world revolution of 1848 broke out in most of the capitals of Europe to challenge monarchies and to promote the self-determination of nations, but it had echoes in the new Christian sects that emerged within the United States and even as far as the Taiping Rebellion in China. The world revolution of 1917 included the Russian revolution, but also the Mexican and Chinese revolutions, the Arab uprising of 1916, and it gave impetus to the great wave of decolonization struggles in Asia and Africa that climaxed after World War II (Goodwin 2001; Foran 2005). In the world revolution of 1968, students mobilized in the United States, Mexico, France, Italy, West Germany, United Kingdom, Poland, Yugoslavia, Japan, Pakistan, Sweden, Spain, Brazil, Jamaica, Uruguay, and China to protest imperialism and racism, but also to protest the perceived failures of the Old Left's project of taking power in nation-states (Gitlin 1993; Chase-Dunn and Lerro 2014).

The New Left of 1968 was a reaction against the perceived failures of the labor movement, the communist parties, and the welfare state. The Port Huron statement, formulated by the Students for a Democratic Society in 1962, was a list of grievances and prescriptions in favor of direct participatory forms of democracy (horizontalism) and resistance to bureaucratization and formal political parties (Flacks 1988).

Horizontalism abjures hierarchy and valorizes egalitarian relations among individuals and groups and consensual decision-making. Horizontalist organization, also called "self-organization" (Prehofer et al. 2005), has several advantages: resilience (you can lose some of them but there is redundancy), flexibility and adaptability, individual entities interact directly with one another, and there is no larger hierarchy that can be disrupted. These desirable characteristics are those that are stressed by advocates of horizontalist networks (Holloway 2002; Juris 2008; Zibechi 2010). But critics of horizontality point out that being structureless does not prevent the emergence of informal structures among groups of friends, and groups that embrace structure have no mechanisms for regulating the power of these informal networks (Freeman 1973).

These ideas were adopted and elaborated by the Global Justice movement and were discussed in chapter 3. Horizontalism is seen as the antidote to organizational hierarchy and bureaucratic sclerosis. The direct participatory decision-making methods employed by the Occupy Wall Street movement, Spanish Indignados, and Argentine popular assemblies reflected many of the same concerns that were expressed in the Port Huron statement. These political principles have stymied the emergence of capacious political organizations such as unions and parties that could mobilize people and contest for power on a large scale. There is a search for new forms of organization that both resist the oligarchical tendencies of organizations and that have the capability of coordinating the activities of large numbers of people for sustained contestation. A new internationalism is emerging that tries to overcome the obstacles that the culture of the Global Left placed in the way of progressive global political organization since the rejection of the Old Left in 1968 (Amin 2019).

In 1989, important movements in the Soviet Union, Eastern Europe, and China challenged Communist regimes in the name of democracy. These movements also challenged activists in the West to reexamine the characterization of civil rights, human rights, and the rule of law as reformist bourgeois (liberal) issues. The rise of twenty-first-century fascism has had a similar effect.

Starting in 1994 with the Zapatista revolt in Southern Mexico, another world revolution has emerged to contest global injustice, continued warfare, autocratic rule, and corporate capitalist austerity policies. A global wave of protests in 2011 erupted when the Arab Spring swept through the Middle East and inspired the indignados anti-austerity protests in Spain and Greece and then the Occupy Wall Street movement (Gitlin 2012; Mason 2013; Curran, Schwarz, and Chase-Dunn 2014). The disappointing outcomes of the mass protests of 2011, along with the demise of the Pink Tide in Latin America, have caused activists of the New Global Left to rethink their theories, analyses, and strategies.[3]

As we have said above, the New Global Left is a component of the larger global civil society of actors who are consciously participating in world politics. Some players within the New Global Left are trying to change the nature of world society, while others are simply trying to defend themselves against larger forces. A significant group is trying to create local communities that are constructed to redress some of the problems that global capitalism has created and that are intended to prefigure a post-capitalist world that is more humane and sustainable. Prefigurationism is the idea that small groups and communities can intentionally organize social relations in ways that can provide the seeds of transformation to a more desirable form of future human society. The New Global Left is just the latest incarnation of a Global Left that has been directly engaging in world politics since the world revolution of 1789.

Each world revolution reflects the nature of contemporary contradictions, the ideological and analytic heritages of earlier world revolutions, and the institutional structures that are predominant during its historical period. World revolutions are complicated because local and national struggles have different and often unique characteristics due to the different histories of each local community and national society, and because people in different zones of the larger system—for example, in the Global North and the Global South—often have different interests and experiences. Nevertheless, each world revolution takes on a character of its own that is due to the nature of the constellation of movements that make it up, the nature of contending movements, and the actions and ideologies of the authorities that are challenged.

As we said in the introduction, global civil society is composed of all the individuals and groups who knowingly orient their political participation toward issues that transcend local and national boundaries and who try to link up with those outside of their own home countries to have an impact on local, national, and global issues. The New Global Left is that subgroup of global civil society that is critical of neoliberal and capitalist globalization, corporate capitalism, and the exploitative and undemocratic structures of global governance (Santos 2006; Steger et al. 2013). The larger global civil society also includes defenders of global capitalism and of the existing institutions of global governance as well as other challengers of the current global order (see chapter 5).

The New Global Left is the current incarnation of a constellation of popular forces, social movements, political parties, and progressive national regimes that have contested with the great powers and global elites since the world revolution of 1789. The existing institutions of global governance have been shaped by the efforts of competing elites to increase their powers and to defend their privileges, but also by the efforts of popular forces and progressive states to challenge the hierarchical institutions, defend workers' rights, access to the commons, the rights of women and minorities, the sovereignty of indigenous peoples (Chase-Dunn et al. 2019) and to democratize the local, national, and global institutions of governance (Smith and Wiest 2012; Chase-Dunn and Lerro 2014).

The New Global Left includes both civil society entities: individuals, social movement organizations, nongovernmental organizations (NGOs), but also political parties, party-networks, and progressive national regimes. In this chapter, we examine the relationships among the movements and the progressive populist regimes that emerged in Latin America as what was called the Pink Tide. These regimes have been an important part of the New Global Left, though the relationships among the movements and the regimes have been both supportive and contentious (Chase-Dunn, Morosin, and Alvarez 2014; Herrera 2014).[4]

The boundaries of the progressive forces that have come together in the New Global Left are fuzzy and the process of inclusion and exclusion is ongoing. The rules of inclusion and exclusion that are contained in the Charter of the World Social Forum, though still debated, have not changed much since their formulation in 2001.[5]

The New Global Left has emerged as resistance to, and a critique of, global capitalism and the neoliberal globalization project (Chase-Dunn 1999). It is a coalition of social movements that includes

- old social movements that emerged in the nineteenth century (labor, anarchism, socialism, communism, feminism, environmentalism, peace, human rights, and national liberation/anti-colonialism) along with
- more recent incarnations of these and the so-called new social movements that emerged in the world revolutions of 1968 and 1989 (queer rights, anti-corporate, anti-neocolonialism, fair trade, sharing economy, global indigenism, as well as

- movements that emerged (and continue to emerge) out of the state sector created under state-led development in the mid-twentieth century;
- even more recent ones such as the Climate Justice movement (discussed in chapter 3), slow food/food rights, autonomism, global justice/alter-globalization, anti-globalization, health/HIV, anti-austerity (discussed in chapter 2), alternative media, open source knowledge, and take back the city.

The explicit focus on the Global South and global justice is somewhat like earlier incarnations of the Global Left, especially the Comintern, the Bandung Conference, the anti-colonial movements, and the anti-imperialist Third Worldist movement in the Global North in the world revolution of 1968 (Prashad 2007). The New Global Left contains remnants and reconfigured elements of earlier Global Lefts, but it is a qualitatively different constellation of forces because

- there are new elements,
- the old movements have been reshaped, and
- a new technology (the Internet) has been used to try to resolve North/South issues within movements and contradictions among movements, as well as providing a significant jump in the size of mobilizations on a global level (Castells 2013; Almeida and Chase-Dunn 2018).

There has also been a learning process in which the earlier successes and failures of the Global Left are being studied to avoid repeating the mistakes of the past—what we call "strategic experience" (Almeida 2014). We observed this movement learning process in the quantitative models presented in chapters 2 and 3. In national-level, anti-neoliberal revolts in Central America, past collective action against neoliberalism predicted current resistance down to the local level. In chapter 3, we demonstrated that the Global Climate Justice movement benefits from past transnational mobilizations over economic justice and anti-war. The relations within the family of anti-systemic movements and among the populist Pink Tide regimes that emerged in Latin America were cooperative, competitive, and contentious. The Global Left must improve its relationship with progressive and reformist regimes if it is to become a serious force in world politics.

Antonio Gramsci's (1971) theorization of the rise of the "modern prince" developed a Marxist analysis of how a new moral and political order could be built that would replace capitalist civilization. In his formulation, organic intellectuals (working activists) would develop a consensual counter-hegemonic civilization and would educate workers and citizens (the subaltern classes) to challenge and overcome the political elites and capitalist owners of Italian society. Gramsci's analysis of ideological hegemony and his project of developing a counter-hegemony has been adapted to the situation of the New Global Left by several thinkers (Gill 2001, 2003; Sanbonmatsu 2004; Carroll 2015, 2016; Carley 2019). This project challenges many of the aspects of the culture of the Global Left that emerged in the world revolution of 1968 and that have been salient elements of the New Global Left. It also challenges many of the main tenets of postmodern philosophy and post-structuralism, intellectual trends that rose in response to the new social movements (Sanbonmatsu 2004). Here, we address two issues: how the contemporary web of popular progressive movements could move in the direction of greater coordination and cooperation amongst themselves—eventuating in global political organization—and how this organization might be able to engage in coalitional politics of the kind that emerged in the united fronts and popular fronts in the twentieth century. We first review some of the sociological literature on coalitions.

Collective Action and Coalition Theories

Many theoretical approaches in social science are relevant for understanding the process of coalition formation. Exchange theory predicts that individuals and groups that actively benefit one another should be more likely to engage in cooperative behavior. Balance theory predicts that "the enemy of my enemy is my friend." Balance of power theory predicts that coalitions in a triad of competing players are most likely to form among the weaker players in opposition to the strongest player. All these theories presume a level of unified rational action that is unlikely to be present when the subjects of analysis are social movements, but they are nevertheless suggestive.

A good summary of the main elements involved in coalition formation among social movements is that by Sidney Tarrow (2005). Regarding coalitions within and between social movements, Tarrow contends that the most common purposes of these are to combat a common threat or to take advantage of an opportunity; hence, the often-temporary nature of coalitions. The common threat or existence of opportunity is what gives rise to the coalition and allows it to exist (see also Van Dyke 2003; Van Dyke and McCammon 2010; Van Dyke and Amos 2017). According to Tarrow (2005), four elements are necessary to maintain a coalition:

1. Members must frame the issue that brings them together with a common interest.

2. Members' trust in each other and believe that their peers have a credible commitment to the common issue(s) and/or goal(s).

3. The coalition must have a mechanism(s) to manage differences in language, orientation, tactics, culture, ideology, etc., between and among the collective's members (especially in transnational coalitions).

4. The shared incentive to participate and, consequently, benefit.

Cooperative action and coalitions vary in intensity and longevity. At one extreme are mergers that involve covenants in which the former parties lose their separate identities and create a new integrated and structured organization. At the other extreme are temporary alliances for specific limited purposes in which the parties maintain their separate identities and organizations.

Transnational social movements have big challenges that more local movements have to a much lesser extent (Dale 2011). There is a global culture in formation (Chase-Dunn 1998; Meyer 2009), but many big cultural differences remain between nations, classes, and ethnic groups. People in different parts of the world-system have different problems and different interests. This is especially true of the Global North and the Global South. Thus, all transnational movements have huge problems of communication and value differences, differences in modes of political expression, differences in the relative importance of issues, and differences in the availability of resources and in what Charles Tilly (2003) called "repertoires of contention"—semi-institutionalized modes of political expression. All these differences make communication difficult and undermine identification and trust. Nevertheless, transnational social movements emerged in earlier centuries when these problems were even more daunting, and yet they managed to form powerful coalitions that were significant players in world politics (Martin 2008). The current availability of less costly technologies of communications and transportation is a great opportunity for organizing movements internationally (Almeida and Chase-Dunn 2018).

Ruth Reitan (2007) usefully elaborates earlier typologies of solidarity among global activists and the people they are trying to influence. McCarthy and Zald (1977) discuss "conscience constituents" who are core supporters of a social movement but do not benefit directly from the accomplishment of that movement's goals. According to Reitan, two forms of solidarity emerge among those distant from the immediate consequences that are the focus of the movement: altruistic solidarity and reciprocal solidarity. Altruistic solidarity occurs when "*sympathy* with the suffering of others who are deemed worthy of one's support seems to be the prevailing affective response among those who choose to act" (Reitan 2007: 51). Altruistic solidarity is characterized by low-risk activism that may be largely apolitical, suppress contentious action, and even reproduce inequality.

On the other hand, "reciprocal solidarity" emerges when "a perceived connection between one's own problems or struggles and that of others tends to lead to empathy with another's suffering and a sense that its source is at least *remotely threatening* to oneself" (Reitan 2007: 51). Reciprocal solidarity is characterized by pluralism and cooperation between conscience constituents and beneficiary constituents in pursuit of structural change. These distinctions, however, are largely analytical, as "movements today are comprised of identity, reciprocal, and altruistic solidarities alike, in different mixes towards different outcomes" (Reitan 2007: 56).

Reitan recognizes the importance and validity of both altruistic and reciprocal solidarity but also considers their limitations. She tells the story of Jubilee 2000, a coalition of churches in the Global North who began a campaign of debt relief for countries in the Global South who had become hugely indebted to banks in the Global North in the last decades of the twentieth century (as discussed in chapter 2). Jubilee 2000 was based mainly on altruistic solidarity with somewhat weak participation from the Global South. But, when the campaign succeeded in bringing banks to the table for negotiations about debt relief, the leadership of Jubilee 2000 made compromises that were seen as a betrayal by the activists from the

Global South, who then formed their own organization, Jubilee South. This story is meant to show the limitations of altruistic solidarity and the necessity for activists from the Global South to have their own autonomous organizations. A related issue is the often-contentious relationships between NGOs (organizations with budgets and paid staff) and social movement organizations that rely on mass memberships and volunteer (unpaid) leadership. Reitan (2007) also tells the story of Via Campesina, a global union of small farmers, that rejected participation by NGOs after these were seen as attempting to steer the organization. Via Campesina opted to restrict membership to farmers only, even excluding friendly participant-observing sociologists as well as NGOs.

More recently Ruth Reitan (2012a: 324) has discussed the "frayed braid" composed of the three strands of the historic Global Left (liberalism, Marxism, and anarcho-autonomism), pointing out that, in the current setting, these strands have not broken apart as they did in previous world revolutions. Reitan contends that the different strands of the braid have learned from the past. She also says, there is "a desire for building inter-movement solidarity and broader alliances while retaining intra-movement identity and autonomy" (Reitan 2012a: 324). Reitan writes that "the process of transnational coalescence entails bridge-builders ... who, when faced with a new bellwether or trigger issue, temporarily scale down (i.e., internalize, or loop back) to the local, national, or macroregional levels, but with the aim to scale up again to the transnational level of contention as a broader movement. To do so, they use various framing tactics while brokering new ties and diffusing information among existing and new activists, particularly frame extension, in order to foster transnational coalescence ... between 'new' and 'old' issues and movements" (Reitan 2012b: 338).

Reitan also discusses the "rooted cosmopolitans" named by Sidney Tarrow (2005: 42). These are activists "whose relations place them beyond their local or national settings without detaching them from locality. Thus, rather than being rootless cosmopolitans, these bridge-builders are among the most committed and seasoned activists, with expertise, leadership experience and ready access to domestic-level material and symbolic resources" (Reitan 2012b: 339). These are the activists featured in Keck and Sikkink's (1998) analysis of "transnational advocacy networks" and they pinpoint one of the motives for upscaling (going transnational). This is the so-called boomerang effect in which social movement activists use their ties abroad to bring pressure on reticent local or national authorities. Immanuel Wallerstein (2004) points to the importance of "synergists" who participate in, and link, different social movements with one another.[6]

Justice Globalism as an Ideological Constellation

Manfred Steger, James Goodman, and Erin K. Wilson (2013) presented the results of a systematic study of the political ideas employed by forty-five NGOs and social movement organizations associated with the International Council of the World Social Forum. Using a modified form of morphological discourse analysis developed by Michael Freeden (2003) for studying political ideologies, Steger, Goodman, and Wilson analyzed texts (web sites, press releases, and declarations) and conducted interviews to examine the key concepts, secondary concepts, and overall coherence of the political ideas expressed by these organizations as proponents of "justice globalism."

The key concepts of justice globalism extracted by Steger et al. (2013: table 2.1, 28–29) are

- participatory democracy,
- transformative rather than incremental change,
- equality of access to resources and opportunities,
- social justice,
- universal human rights,
- global solidarity among workers, farmers and marginalized peoples, and
- ecological sustainability.

More detailed meanings of each of these concepts have emerged in an on-going dialectical struggle with market globalism (neoliberalism). Steger et al. discuss each of these and evaluate how much consensus exists across the forty-five movement organizations they studied. They find a large degree of consensus, but their results also reveal a lot of on-going contestation among the activists in these organizations regarding the definitions and applications of these concepts.

For example, though most of the organizations seem to favor one or another form of participatory democracy, there was awareness of some of the problems produced by an overemphasis on horizontalist processes of participation and ongoing debates about forms of representation and delegation.

Some of the organizations studied by Steger et al. eschew participation in established electoral processes, while others do not. Steger et al. highlight the importance of "multiplicity" as an approach that values diversity rather than trying to find "one size fits all" solutions. They note that the Charter of the World Social Forum values inclusivity and the welcoming and empowerment of marginalized groups. Prefiguration has found wide support from most global justice activists and social movement organizations. The Zapatistas, the Occupy Wall Street activists, and many in the environmental movement have engaged in efforts to construct more egalitarian and sustainable local institutions and communities rather than mounting organized challenges to the global and national structures of power. The discussion of global solidarity in Steger et al. emphasizes the centrality of what Ruth Reitan (2007) has called "altruistic solidarity"— identification with poor and marginalized peoples—without much consideration of solidarity based on common circumstances or identities. Steger et al. do, however, mention the important efforts to link groups that are operating at both local and global levels of contention.[7]

The Steger et al. study is a useful example of how to do research on political ideology and it provides valuable evidence about ideational stances and culture of the New Global Left. It and the movement network results summarized below imply that the New Global Left has a degree of coherence that can be the basis of greater articulation.

Transnational Alternative Policy Think Tanks

William Carroll's (2016) thorough study of alternative policy groups examined the problem of how to build a transnational counter-hegemonic bloc of progressive social forces (Carroll 2016: 23). Carroll's study examined sixteen progressive transnational think tanks from both the Global North and the Global South.[8] Carroll's results agree with the findings of the Steger et al. (2013) study summarized above regarding the discursive content of the Global Justice movement. Carroll also notes that the progressive counter-hegemonic think tanks that he has studied have been trying to produce knowledge that is useful for prefigurative social change and a democratic and egalitarian forms of globalization in contrast to the neoliberal globalization project. Carroll critiques localist and anti-organizational approaches and proposes a counter-hegemonic "globally organized project of transformation aimed at replacing the dominant global regime with one that maximizes democratic political control and makes the equitable development of human capabilities and environmental stewardship its priorities" (Carroll 2016: 30). We have been fortunate to have a global forum, the World Social Forum process, as a central venue for studying the New Global Left, but some important progressive social movements have been excluded or have excluded themselves from the World Social Forum process. Nevertheless, the Social Forums (global, national, and local) have provided convenient opportunities for studying progressive activists, but how representative these are of the progressive forces in world politics remains an important question.

The World Social Forum Surveys

Our research on the World Social Forum produced maps of the network of movements that are involved in the Social Forum process (see figure 4.1 below). This is a representation of the structure of the left-wing portion of global civil society. The University of California-Riverside Transnational Social Movement Research Working Group conducted four paper surveys of attendees at Social Forum events.[9]

We used previous studies of the Global Justice movement by Amory Starr (2000) and by William Fisher and Thomas Ponniah (2003) to construct our original list of eighteen social movement themes that we believed would be represented at the January 2005 World Social Forum in Porto Alegre, Brazil. We also conducted a survey at the WSF in Nairobi, Kenya, in 2007 in which we used most of these same movement themes, but we separated human rights from anti-racism and we added eight additional movement themes (development, landless, immigrant, religious, housing, jobless, open source, and autonomous). We used this same larger list of twenty-seven movement themes at the US Social Forum (USSF) in Atlanta in 2007 and in Detroit at the USSF in 2010.

We studied changes and continuities in the relative sizes of movements and changes in the network centrality of movements. Relative movement size was indicated by the percentage of surveyed attendees who claimed to be actively involved in each movement theme. We asked each attendee to check whether they identified with, or were actively involved in, each of the movement themes with the following items on our survey questionnaire:[10]

Check all the following movements with which you (a) strongly identify with and/or (b) are actively involved in:

(a) *strongly identify*:	(b) *are actively involved in*:
1. ☐ Alternative media/culture	☐ Alternative media/culture
2. ☐ Anarchist	☐ Anarchist
3. ☐ Anti-corporate	☐ Anti-corporate
4. ☐ Anti-globalization	☐ Anti-globalization
5. ☐ Anti-racism	☐ Anti-racism
6. ☐ Alternative globalization/ global justice	☐ Alternative globalization/ global justice
7. ☐ Autonomous	☐ Autonomous
8. ☐ Communist	☐ Communist
9. ☐ Development aid/ economic development	☐ Development aid/ economic development
10. ☐ Environmental	☐ Environmental
11. ☐ Fair trade/trade justice	☐ Fair trade/trade justice
12. ☐ Food rights/slow food	☐ Food rights/slow food
13. ☐ LGBTQ rights	☐ LGBTQ rights
14. ☐ Health/HIV	☐ Health/HIV
15. ☐ Housing rights/anti-eviction/squatters	☐ Housing rights/anti-eviction/squatters
16. ☐ Human rights	☐ Human rights
17. ☐ Indigenous	☐ Indigenous
18. ☐ Jobless workers/welfare rights	☐ Jobless workers/welfare rights
19. ☐ Labor	☐ Labor
20. ☐ Migrant/immigrant rights	☐ Migrant/immigrant rights

21. ☐	National sovereignty/ national liberation	☐	National sovereignty/ national liberation
22. ☐	Open source/intellectual property rights	☐	Open source/intellectual property rights
23. ☐	Peace/anti-war	☐	Peace/anti-war
24. ☐	Peasant/farmers/landless/ land reform	☐	Peasant/farmers/landless/ land reform
25. ☐	Religious/spiritual	☐	Religious/spiritual
26. ☐	Socialist	☐	Socialist
27. ☐	Women's/feminist	☐	Women's/feminist
28. ☐	Other(s), Please list	☐	Other(s), Please list

The structure of the network of social movement themes that we found when we combined the results of our three surveys in Nairobi, Atlanta, and Detroit (those surveys in which we had twenty-seven movement themes) is shown in figure 4.1. Those movements near the center of the diagram are more central in the matrix of movement links, meaning that they were bigger and had more connections with other movements than those further from the center. Those out on the edge are less central in the sense that they are smaller and share overlaps with fewer other movements. The list in the upper left-hand corner was not very connected with the other movements. But the actual matrix of movement connection had no zeros, so all the movements were connected to at least some other movements by overlapping participants. The outlying movement themes in the upper left-hand corner of figure 4.1 seem unconnected in the figure because in order to perform network analysis it is necessary to dichotomize valued data into 0s and 1s. Those in the upper left-hand corner were below the cutting point of this dichotomization, but they were still connected by some overlaps with movements in the main cluster. The central cluster of movement themes to which all the other movements were linked included human rights, anti-racism, environmentalism, feminism, peace/anti-war, anti-corporate, and alternative globalization.

Many observers of the Global Justice movement somewhat overemphasize the extent to which the movement has been incoherent regarding goals ("one no, many yeses") and shared perspectives. Our surveys of attendees at both world-level and national-level Social Forums have found a relatively stable multicentric network of movement themes in which a set of more central movements serve as links to all the other movements based on the reported active involvement of activists with movements (Chase-Dunn and Kaneshiro 2009). All the movement themes used in the surveys were connected to the larger network by means of co-activism, so it was a single linked network without sub-cliques. This multicentric network was quite stable across venues and over time.[11] This suggests that there has been a similar structure of network connections among movements that are global in scope and that the global-level network is also very similar to the network that exists among Social Forum activists from grassroots movements within the United States (Chase-Dunn and Kaneshiro 2009). We also found that, though both anarchists and indigenous rights activists were few among the attendees at the Social Forums, many of the political ideas of both anarchists and *indigenistas* were widely supported (Aldecoa et al. 2019; Chase-Dunn et al. 2019). The linked network of movements we found in the Social Forum process could be the basis of a more effective global organizational instrument, especially if it could coalesce around the threat of anthropogenic climate change.

Fronts and Coalitions

The history of broad-based, left-wing movement coalitions in earlier periods is relevant for understanding articulation processes in the contemporary world revolution. The Third International (the Communist International or Comintern) was a complex of red networks assembled to coordinate the political actions of communists in the years after World War I. A transnational group of communist intellectuals and organizers claimed to lead the global proletariat in a world revolution that was intended to transform capitalism into socialism and communism by abolishing large-scale private property in the means of production (Hobsbawm 1994).

The Comintern adopted its own statutes at its second congress in 1920. It was led by an executive committee and a presidium. The statutes stated that congresses with representatives from all over the world were to meet "not less than once a year." The Comintern also organized and sponsored several other "front organizations"—the Red International of

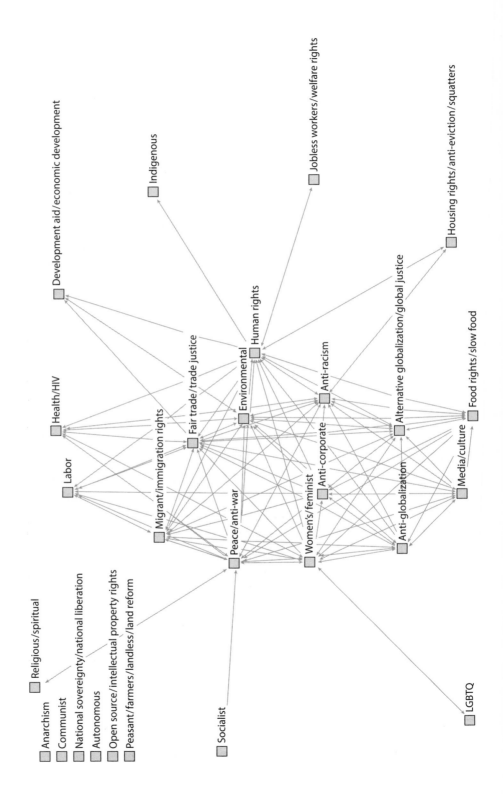

Anarchism

Communist

National sovereignty/national liberation

Autonomous

Open source/intellectual property rights

Peasant/farmers/landless/land reform

Religious/spiritual

Development aid/economic development

Indigenous

Jobless workers/welfare rights

Housing rights/anti-eviction/squatters

Health/HIV

Human rights

Fair trade/trade justice

Anti-racism

Alternative globalization/global justice

Labor

Environmental

Migrant/immigration rights

Anti-corporate

Food rights/slow food

Peace/anti-war

Women's/feminist

Anti-globalization

Media/culture

Socialist

LGBTQ

Labor Unions, the Communist Youth International, International Red Aid, the International Peasants' Council, the Workers' International Relief, and the Communist Women's Organization (Sworakowski 1965; Comintern, n.d.).

The Comintern was founded in the Soviet Union, the "fatherland of the proletariat," and so it is often depicted as having been mainly a tool of Soviet foreign policy. There is little doubt that this became true after the rise of Stalin. In perhaps the most blatant example, Stalin tried to use the Comintern to get communist parties all over the world to support the Hitler-Stalin pact of 1939. But during Lenin's time, the Comintern held large multinational congresses attended by people who spoke at least forty different languages as their native tongues. The largest of these congresses had as many as sixteen thousand delegates attending. Sworakowski (1965: 9) says, "After some attempts at restrictions in the beginning, delegates were permitted to use at the meetings any language they chose. Their speeches were translated into Russian, German, French and English, or digests in these languages were read to the congresses immediately following the speech in another language. Whether a speech was translated verbatim or digested to longer or shorter versions depended upon the importance of the speaker. Only by realizing these time-consuming translation and digesting procedures does it become understandable why some congresses lasted as long as forty-five days."

The Comintern was abolished in 1943, though the Soviet Union continued to pose as the protagonist of the world working class until its demise in 1989. Paul Mason (2013) reminds us of the importance of threats from other social movements that pose challenges that drive former sectarians to try to be more inclusive and pragmatic. The United Front originated as an effort by Communists to create an alliance with other socialists, peasants, and all workers. The Popular Front was an even broader coalition that included all those who were willing to oppose fascism, including capitalists and their parties. These efforts have usually been seen as manipulative moves by communists to infiltrate and control other movement organizations, but David Blaazer's (1992) study of Popular

FIGURE 4.1 (OPPOSITE)   Movement Links: Number of Affiliations Based on Reported Active Involvement in Twenty-Seven Movement Themes from the Social Forums Surveys in Porto Alegre, Nairobi, Atlanta, and Detroit.

Front leaders in Britain shows that many of the noncommunist participants were not ignorant dupes of the communists. They were committed
democrats and socialists who were willing to work with communists to
mobilize the fight against fascism (see also the essays in Graham and Preston, 1987). The Comintern also assisted in uniting rural and urban labor
in the Global South against fascism and state repression. For example,
the International Red Aid organization was active in El Salvador in the
early 1930s (Gould and Lauria-Santiago 2008). The generalized reciprocity of the Left exhibited by the united and popular fronts of the 1930s
lived on in the World Revolution of 1968 during its broad movement
phase. The New Left developed a critique of the failed institutions of the
Old Left (unions and political parties) but many New Leftists displayed
generalized reciprocity in their willingness to work in coalition with Old
Leftists for purposes of ending the war in Vietnam and pursuing a world
revolution of the young (Gitlin 1993). The "unfrayed" braid observed by
Reitan in the Social Forum process had earlier incarnations.

But the general point made by Mason stands. The forces of divergence in the Global Left of the 1930s were partly overcome by the clear
and present danger of the rise of a great wave of fascism (see Goldfrank
1978 and chapter 5 of this volume). Of course, during the midst of this
wave of cooperation the Trotskyists proclaimed a Fourth International in
1938 (James 1997). And after the demise of fascism that resulted from the
outcome of World War II, the left fragmented into sectarian groups again
in most countries. Maoists in China managed to put together a coalition
strong enough to beat the Nationalists and to proclaim the Peoples Republic. But in other countries, new fissures emerged between different kinds
of Maoists and other factions of the radical left.

What are the implications of this history of fronts and fissures for the
near-future situation of the Global Left? Will twenty-first century neofascism and right-wing nationalist populism drive the different elements of
the New Global Left together? Counter-movements are movements that
emerge to counteract and oppose the efforts of other movements (Snow
and Soule 2010: 82). William I. Robinson (2013) contends that recent developments in response to the growth of resistance from below and the
disarray caused by the various contemporary crises of global corporate
capitalism constitute the rise of "twenty-first-century fascism" in the guise
of a globally coordinated police state. Shoshana Zuboff (2019) has de-

scribed the rise of "surveillance capitalism." Such a development could become such a large threat that the divergent elements in the New Global Left might be forced to forge a strong and organized response. The identity movements, the horizontalists and the prefigurers would need to overlook some of their differences to put together a more pragmatic and better organized instrument with which to counter twenty-first-century fascism.

The issue of coalitions and the balance theory notion that the enemy of my enemy is my friend suggest the importance of a distinction between progressive and counter hegemonic. Jihadist Islam and some versions of other religious fundamentalisms are counter-hegemonic in that they are rebellions against the institutions and cultural tropes of the neoliberal globalization project. In chapter 5, we contend that these religious fundamentalisms, despite being protests against global capitalism, are part of the Global Right and should not be considered allies by the Global Left because of their stances on gender inequality and other issues.

States and Social Movements

Our concern for capability in world politics requires attention to the contentious relationship between progressive social movements and the progressive populist Pink Tide regimes that emerged in many Latin American countries in the late twentieth and early twenty-first century. We agree with Patrick Bond (2013) that many of the semiperipheral state challengers to the hegemony and policies of the United States (the so-called BRICS) seem mainly to be trying to move up the food chain within the capitalist world-system rather than trying to produce a more democratic and sustainable world society.[12] Revolutions are needed within these polities to produce regimes that will be effective agents of transformative progressive social change. This said, transnational social movements should be prepared to work with progressive regimes that emerge in order to try to change the rules of the global political and economic order (Evans 2009, 2010, 2020).

It is well-known that many transnational movement organizations scorn politics-as-usual and resist efforts by progressive regimes to provide resources and leadership to movements. Autonomism makes this a basic principle and the World Social Forum Charter proscribes individuals from attending as representatives of governments. This is part of the anti-elitism of the culture of progressive grassroots movements. In 2005, Presidents Lula of Brazil and Chávez of Venezuela had to give their speeches at a venue near to, but not part of, the World Social Forum meetings in Brazil.[13] When nineteen prominent leftist academics tried to issue a declaration in the name of the World Social Forum at end of the meeting in Porto Alegre in 2005, they were widely denounced as elitists. This is related to the horizontalist stance that has been strong in the Social Forum process since its emergence. Most activists either want no leaders or want leaders to "bubble up from below." The Zapatistas of Chiapas took this position. Similar stances were widespread in the student movements of the 1960s (Gitlin 1993). Facilitators were preferred over charismatic grandstanders. The New Left critique of the Old Left was heavily based on a rejection of the goal of taking state power, which had become an end in itself rather than a means to the transformation to a more just and equal society. The New Leftists and the global justice activists are strong believers in what Robert Michels (1962) termed "the oligarchical tendencies of political parties."

We contend that the anti-organizational ideologies that have been salient elements of the culture of progressive movements since 1968 have been a major fetter restricting the capability of progressive movements to effectively realize their own goals. But these ideas and sentiments run deep and so any effort to construct organizational forms that can facilitate progressive collective action must be cognizant of this embedded culture. At the same time, political parties, religious institutions, and portions of the state sector (such as health and education) maintain the most extensive territorial and organizational reach in the current neoliberal period (see chapter 2). The internet and social media, allowing cheap and effective mass communication, have been blamed for producing specialized single-issue movements. We suggest that digital communication can be harnessed to produce more sustained and integrated organizations and effective tools that can be used to contend for power in the institutional halls of the world-system. We also think that the old reformist/revolutionary debate about whether to engage in electoral politics is a fetter on the ability of the Global Left to effectively contend. States are not, and have never been, whole systems. They are organizations, like the Boy Scouts of America. And their organizational resources can be used to facilitate the building of a post-capitalist global society. Progressive transnational social movements should be prepared to work with progressive regimes in order to try to change the rules of the global political and economic order (Evans 2009, 2010).

The "leaderless" discourse of the Occupy Wall Street movement was another incarnation of horizontalism. Horizontalism is constituted by a strong commitment to the value of each unique individual, and the equality of individuals, and to the empowerment of marginalized groups. In this sense it is redolent of the global moral order described and analyzed by John W. Meyer (2009). Radical individualism has been an important feature of many millenarian religious and political movements since medieval times (Cohn 1970).

The comparative world historical perspective on state/movement relations stresses the importance of the effects that earlier world revolutions had because of the emergence of regimes in the semiperiphery that explicitly challenged the existing world order and the global rule of capital.[14] The rise of the Bolshevik Regime in Russia, as well as strong labor movements within the core states, spurred the New Deal and social democracy to save capitalism by reforming it. In the world-systems perspective states are organizations that claim sovereignty but that are interdependent parts of a larger political economy—the interstate system and the world economy. A movement that attains state power in a modern national state has not conquered the whole system. It has taken over an institution that is part of the larger system.

This explains much about the policies of the so-called communist states that came to power in the twentieth century. They engaged in semiperipheral protectionism in order to industrialize and they invested huge resources on military capability in order to prevent conquest by capitalist core states. It is entirely understandable why social movements should seek to maintain their autonomy from states, but the principle of nonparticipation in "politics-as-usual" should be replaced by recognition of the functionality of coordination between radical and reformist social movement organizations. Radical movements that threaten to transform the whole social order increase the likelihood that enlightened conservatives will make deals with less radical and more legitimate social movement organizations and NGOs.[15] This is how it has worked in the past and how it is likely to work in the future. Awareness of this dynamic should be useful to social movement organizations, promoting greater tolerance and collaboration between radicals and reformists.

The Global Class Structure

Analysts of the alleged global stage of capitalism contend that a transnational capitalist class has recently emerged, and that global capitalism is also producing a newly transnationalized working class (e.g., Sklair 2001; Robinson 2008, Robinson 2019). This analysis has the implication that class struggle should now be occurring at the global level. World-systems analysts have been studying the system-wide configuration of classes over the past several hundred years (Wallerstein 1974; Amin 1980b) but whether or not recent changes are seen in long-run perspective, there have obviously been important recent changes, and these have implications for analyzing potential movements that might emerge in response to recent crises. The analysts of global capitalism have talked about "the peripheralization of the core" as a way of describing the attack on core workers and unions that has been carried out by neoliberals (Ross and Trachte 1990; Moody 1997). The casualization of labor and the growth of the informal sector has been an important phenomenon in both core and noncore countries since the rise of Reaganism/Thatcherism (Standing 2011). Beverly Silver's (2003) study of waves of labor unrest shows that the export of the industrial proletariat from the core to the semiperiphery produced militant labor movements in the new regions of industrial manufacturing in the semiperiphery.

Guy Standing (2011) argues that the old industrial proletariat of workers in large Fordist factories has been eclipsed by the rise of a new class, the *precariat* of workers of the gig economy with precarious employment situations. Standing describes in detail the political, cultural, and economic processes that have produced the precariat. He sees different layers of the precariat in the overall global class structure. The main division within the precariat is between middle-class, educated youth that cannot get, and do not want, full-time jobs and the unemployed poor. The latter are especially prevalent in the cities of the Global South.

Standing sees the old working class as shrinking, but he does not pay much attention to the movement of formal wage jobs in factories to the semiperiphery, especially China. He proposes that the different fractions of the precariat should come together to challenge neoliberal capitalism, but he also recognizes that this is difficult because the middle-class millennials and the urban poor have different cultures and perceive themselves to have different interests (Standing 2014). Standing's vision of a movement to defend and rebuild workers' rights became more plausible in the 2016 US presidential campaign of Bernie Sanders and the democrats running against Trump in the 2020 election. But the election of workers' rights candidates would have to be accompanied by the kind of mass mobilization that Sanders and others have been advocating in order to become a reality.

Savan Savas Karataşli, Sefika Kumral, Ben Scully, and Smriti Upadhyay (2014) studied the wave of unrest that spread around the globe from 2008 to 2011, building upon Silver's (2003) studies of labor unrest. Using protest and labor unrest data coded from major news sources, they provided a useful review of efforts in the social science literature to analyze this global cycle of protest. They noted that most analyses underplay the significance of the role of wage-earners in these protests. They used their coded protest data to examine the extent to which this recent wave of protests indicates a recurrence of past forms of unrest or whether the current period represents a different pattern.

Silver (2003) divided labor protest into two categories: Marx-type and Polanyi-type. Marx-type unrest refers to offensive struggles of emergent working classes, whereas Polanyi-type unrest refers to the defensive protests of workers whose previous gains were being undermined as well as resistance against proletarianization. Karatasli et al. (2014) find that, along with a mix of Marx-type unrest[16] and Polanyi-type unrest,[17] which are part of the older cyclical process of the making and unmaking livelihoods by capitalism,[18] a third type of unrest, driven by what Marx called *stagnant relative surplus population*,[19] also played a large part in the protests of 2011 and presents a "secular trend in which capitalism destroys more livelihoods than it creates over time." The growing number of this third type of worker, composed largely of those who can work but are unable to be absorbed by the productive capacity of the economy, and their presence in protests and social movements that have continued in the aftermath of 2011 (see also figures 2.1 and 2.2 in chapter 2), gives credence to the idea that the problems faced in regions of unrest are chronic and enduring. This growing excluded segment of the population is an important force in these movements, and the growing number of redundant potential workers is reflected in the huge waves of immigration that have spurred the anti-immigrant politics of the New Global Right (Mora et al. 2017; Mora et al. 2018).[20]

Karatasli et al. (2014) criticize those who focus discussion on the precariat (e.g., Standing 2011; see also Korotayev and Zinkina 2011) of middle-class workers whose job conditions and incomes have declined and on educated young people who cannot get a job commensurate with their expectations and face large education-related debts ("graduates without a future").[21] Both these and the less educated young who cannot find employment can be understood as different parts of the stagnant relative surplus population. Karatasli et al. (2014) strongly demonstrate that workers have played a large role in the 2011 protest wave, and they correctly note that this has often been overlooked by other analysts of these protests.

Contenders for Articulation

Which of the existing transnational social movements and movement coalitions that have come out of the Social Forum process could plausibly emerge as central to the formation of a more capable coalition of the New Global Left that could strongly challenge the global rule of capital in the twenty-first century? We agree with Steger et al. that a coherent ideological framework already exists. Our Social Forum survey findings support the optimism of Ruth Reitan and Jackie Smith (Smith and Wiest 2012) regarding the existence of an already functioning collaboration within the Social Forum process and elsewhere. But we also see the need for a stronger and more capable instrument to play a consequential role in world politics in the emerging period of crises. What could be the pivot of such a new articulation?

Workers (Again)

Could a reconfigured anti-corporate movement based on workers' rights and global unionism come out of the global precariat that has been produced by neoliberal capitalist attacks on labor unions and the welfare state? Peter Waterman (2006) proposed a Global Labor Charter that was intended to mobilize such a coalition and Guy Standing (2014) proposed a charter for the precariat. Mike Davis (2006) has suggested that the informal sector workers of the Global South might step forward as an historical subject. And William I. Robinson (2008, 2019) has theorized the emergence of a transnational working class that has been created by the processes of global capitalism. Robinson (2019: 72–73) says, "The hope of humanity now lies with a measure of transnational social governance over the process of global production and reproduction." He proposes that this democratization of global society should be led by the new global proletariat.

Austerity politics by neoliberals would seem to provide the basis for such a movement, and elements of an anti-austerity coalition made some headway in the European Summer and the Occupy movements and several of the national-level uprisings discussed in chapter 2. Perhaps a reconfigured version of the Old Left notion of the world working class as the midwife of a less oppressive human future might yet be able to articulate the anti-systemic movements. Social movement unionism and experiments in cross-border organizing have had some successes, as well as notable failures (Armbruster-Sandoval 2005; Anner 2011); but in a deepening crisis these efforts might do the job.

The precariat also includes the unemployed educated seen by Mason as the main participants in the recent waves of popular protest demonstrations that have emerged since the Arab Spring. The educated unemployed (or underemployed) are also burdened with large debts incurred when neoliberals shifted much of the cost of public higher education on to students. In the context of growing levels of income and wealth inequality within many core countries (Picketty 2014) this would seem to prepare the ground for social movements in which radicalized middle class elements might once again ally with the urban poor and workers, as they did in the world revolution of 1848 (Mason 2013).

Feminists

Maria Mies (1986) argued that women and the marginalized peasants and workers of the Global South formed an exploited and potentially revolutionary subject that could rise to challenge global capitalism. Ecofeminists have emphasized the complementarities of a kinder, gentler approach to nature and the politics of women. Socialist feminists have noted the growing importance of female labor in all the world's regions and the leadership shown by global feminists in confronting, and partially resolving, North/South differences among women (Moghadam 2005). Anarchists such as David Graeber (2013) have noted that many of the processual innovations that were utilized in the Occupy movement came out of feminist political practices. Feminists have links with many of the other movements, as shown by the results of our research on the network of social movement connections in figure 4.1.

Climate Justice

Arguably the most imminent crisis produced by contemporary global capitalism is the onrushing arrival of anthropogenic climate change. The environmental movement has strong links with some elements of the labor movement, with global indigenism and with feminism (see figure 4.1 above). Patrick Bond (2012) has written convincingly of the emerging centrality of the Climate Justice movement. Climate justice emerged from the environmental justice movement, which was a combination of environmentalism and human rights and anti-racism (Kaneshiro et al. 2015). It has long been noted that the poor are the first to suffer the effects of pollution and environmental degradation. Steger et al. (2012) devoted an entire chapter of their study of justice globalism policy implications to the "climate crisis" and Ruth Reitan and Shannon Gibson (2012) studied three climate activist networks that participated in the Copenhagen climate summit in 2009, supporting the notion that climate justice has great potential as a unifying framework for the New Global Left (see also Hadden 2015). Chapter 3 explained the rapid diffusion of climate justice actions across world regions between 2005 and 2019. Climate justice provides the most elaborate global infrastructure to launch new rounds of transnational collective action in the 2020s.

Post-capitalism

Paul Mason (2015) has outlined a thoughtful and provocative proposal for a transition to post-capitalism—a cooperative decentralized and sustainable global sharing network in which the goods necessary for human life are free, or nearly free, and people can spend their time in creative network projects. Mason's big idea is that information technology makes communism possible because it is nearly costless to replicate and share software and the expansion of information sharing contradicts both markets and private property. This will transform the logic of capitalism into a sharing network. Rather than the rise of the new globalized working class, Mason proposes the abolition of work. The kind of working-class culture and solidarity still romanticized by workerists (e.g., Davis 2018) has been made impossible by precarity and "the freedom to tweet."[22] Work and leisure time have blurred together in the culture of Silicon Valley. The protagonists of post-capitalism are networked individuals. Struggle has moved from the workplace to "society." This proposal seems to be directed mainly to computer programmers and other information technologists of the Global North and to be inspired by the millennial precariat that Mason studied in his close ethnographies of the mass protests that peaked in 2011 (Mason 2013). While Mason is right that the Old Left unfairly vilified utopian socialism, and while his discussion of a possible transition to post-capitalism displays a realistic awareness of the huge demographic and environmental problems that will be chronic in the twenty-first century, the real prospects for unskilled workers of the both the Global North and the Global South in post-capitalism are less than clear. What is needed for the next decades is a post-capitalism that meets the needs of both techies and people who do not have high information technology skills.

Metahumanism

John Sanbonmatsu's (2004) argument for a postmodern prince is somewhat similar to Paul Mason's with regard to the focus on agency. Whereas Mason sees the agents of post-capitalism as creative individuals, Sanbonmatsu's version of humanism focuses on empathy and love. Sanbonmatsu nails postmodern and post-structural philosophy as convenient ways for academic critical theorists to be radical without being Marxist. He notes the symbiosis between the turn toward a focus on language and culturally constructed categories as the sources of oppression with the rise of flexible specialization and precarious labor in the Global North. Sanbonmatsu provides a compelling rendition of Antonio Gramsci's struggle to produce a counter-hegemonic civilization and socialist moral order for the Italian proletariat. Sonbonmatsu contends that a new version of this project should be developed in the twenty-first century. The postmodern prince is a new, or renewed, humanistic moral order based on empathy and love. Its protagonists are committed organic intellectuals who are willing to come together and to lead a global social movement. Sanbonmatsu also advocates that this moral order should include other non-human sentient beings, by which he means animals who can suffer and perceive. He is not afraid of the similarity of his secular metahumanism with Christianity. He sees the job of the postmodern prince as the proclamation and collective production of a new moral order. He also defends Jacobinism, and by extension, Leninism, though his criticism of Lenin is thoughtful. Metahumanism solves the problem of combining the interest of workers with the interests of those who are oppressed by cultural categories and language—including racism, gender hierarchy, discrimination based on sexual preference, and other forms of oppression that have been the province of identity movements (and it extends citizenship to sentient animals). In this regard, it is an expanded version of what has become known as social movement unionism. The intersectionality framework developed by critical race scholars on multiple forms of oppression (Chun et al. 2013; Luna 2016; Terriquez et al. 2018) have now been appropriated by activists to produce massive rainbow coalitions as witnessed in the 2017 and 2018 anti-Trump Women's Marches. These intersectionality perspectives offer the possibility of a coalitional umbrella for overcoming long-standing divisions across racial, ethnic, class, gender, sexuality, citizenship, and Global North/South lines.

Of course, there are other possibilities for the role of articulator. We have mentioned Ruth Reitan's (2012b) discussion of the possible unification of the peace movement and the Global Justice movement. Several "new internationalisms" have been proposed recently. In November of 2018, Bernie Sanders and Yanis Varoufakis issued a call for a Progressive International to unite against the rise of neofascist and right-wing populist parties (Progressive International 2018). Samir Amin (2019) called for the establishment of a fifth international that would coordinate progressive movements and regimes in the Global South with progressives in the Global North. Twenty essays addressing Amin's proposal have been co-published by *Globalizations*, the *Pambazuka News*, and the *Journal of World-Systems Research* (Gills and Chase-Dunn 2019). Àlvarez and

Chase-Dunn (2019) proposed the building of a diagonal political organization for the Global Left that would link local and national networks and prefigurational communities[23] to contend for power in the world-system during the next few decades of the twenty-first century. They discuss the reasons why the Global Left can no longer be content with its amorphous structure as a movement of movements. They contend that the Social Forum process needs to be reinvented for the current period of rising neofascist and populist reactionary nationalism and to foster the emergence of a capable instrument that can confront and contend with the global power structure of world capitalism and the popular reactionary movements that have emerged. This will involve overcoming the fragmentation of progressive movements and identity politics that have been one outcome of the rise of possessive individualism, the internet, and social media. The proposed vessel for the Global Left would focus on struggles for human rights, climate justice, anti-racism, feminism, sharing networks, peace alliances, taking back the city, progressive nationalism, and confronting and defeating neofascism.

Capitalism is in crisis again and the forces of progress are once again moving to try to create a more humane, democratic and sustainable world society. The order, speed of onset, and interaction of different kinds of crises emerging the twenty-first century will favor either a neofascist police state, a reformist global Keynesianism or a more radical and progressive restructuring of political and economic institutions. A coherent social science perspective exists with which to analyze the structures and institutions of the system (the comparative evolutionary world-systems perspective) and a coherent political ideology, justice globalism, has come out of the Social Forum process. What is needed now is organization. Climate justice, feminism, a new version of the workers' movement, intersectional alliances, and metahumanism are possible frameworks for uniting those who want to build a new society within the skeleton of the old with those who want to reorganize the whole system. That will require a capable organizational instrument and the willingness to unite.

FIVE

Evolution of the Global Right in the Geoculture

....

This chapter examines the roots of contemporary right-wing populism and neofascism in the context of world historical change over the past two centuries. It describes the ideological structure of the global political geo-culture and the changing nature of the Global Right by comparing twentieth-century fascism with the neofascist and authoritarian populisms that have emerged in recent decades. World historical waves of right-wing populism and fascism have been caused by cycles of globalization and de-globalization, the rise and fall of hegemonic core powers, long business cycles (the Kondratieff wave), and interactions with both Centrist Liber-alism and the Global Left. We consider how crises of the global capitalist system have produced right-wing backlashes in the past, and how a future terminal crisis of capitalism might lead to a reemergence of a new form of the tributary mode of accumulation or a democratic global polity.

Just as there has been a Global Left since the emergence of the ideo-logical hegemony of Centrist Liberalism after the French Revolution, there has also been a Global Right.[1] The three camps together constitute the ideological constellation of world politics that Immanuel Wallerstein (2012) has called the "geoculture." As we have explained in the Introduc-tion, most world-systems have been multicultural systemic networks. The contemporary system is still substantially multicultural, but the national and civilizational cultures have converged and increasingly form an inter-active and structured whole—global culture. The political geoculture is part of a set of complicated and interacting broadly held, but not entirely predominant, institutionalized assumptions about what exists (ontology) and what is good (a moral order). This broader global culture includes more than just political beliefs and institutions. It includes the predomi-nant ontologies of the universe and of life, the nature of time and beliefs about human nature. Science, humanism, and formal rationality are the key tropes (Meyer 2009).

Immanuel Wallerstein (2012) analyzed the rise of the political geoculture in the nineteenth century. Since the French Revolution, there have been three interacting factions: a progressive camp that favors egalitarianism and democracy, a reactionary faction that favors traditional hierarchies and institutions, and a pragmatic Centrist Liberal faction that seeks stability by combining strategic coercion with efforts to engineer consensus by making partial compromises with popular movements and political forces of both the left and the right.

All three camps—the Global Right, the Global Left, and Centrist Liberalism—have coevolved since the nineteenth century in the context of major changes in the structure of the world-system and their effects on each other. Noël and Thérien (2008: 14–16) describe the emergence of the left-right distinction in the French General Assembly and its reconstitution in the last years of the nineteenth century with the rise of the labor movement and socialism. They do not include a middle group of compromisers, as does Wallerstein, but Centrist Liberalism has been an important recurring feature of world politics in the nineteenth, twentieth, and twenty-first centuries. Noël and Thérien (2008) note that the Global Right shifted from defending monarchy and the church to emphasizing equality of individual opportunity (meritocracy) in a context of a non-interventionist state in response to the rise of socialism in the late nineteenth and early twentieth centuries. Centrist Liberalism also defends property by insisting upon a strong separation between economic and political rights in the theory of polyarchy that recognizes political equality before the law but refuses to contemplate economic equality and economic democracy (Robinson 1996).

The structural context of world politics over the past two centuries has seen successive waves of global economic integration and deglobalization (Chase-Dunn 1999; Chase-Dunn, Kawano, and Brewer 2000); but also the rise and fall of hegemonic core powers and forty-to-sixty-year phases of economic growth and stagnation (Kondratieff waves).[2] And the constellation of states has changed greatly. The rise and fall of British hegemony, the rise of Germany and Japan as semiperipheral challengers, the rise and eventual stagnation of the economic hegemony of the United States (Chase-Dunn, Kwon, Lawrence, and Inoue 2011), the decolonization of the colonial empires and the emergence of "new nations" in the Americas, Asia, Oceania, and Africa have greatly changed the structure that is the arena of world politics over the past two centuries.[3] The main issue we are addressing in this chapter involves the similarities and differences between the successive waves of the Global Right.

Evolution of the Geoculture: Right, Center, and Left

Centrist Liberalism evolved with the changing nature of predominant capitalism, adapting to the structural cycles and reorganizations listed above and in reaction to the changing ideological and organizational aspects of both the Global Right and the Global Left. There have been "varieties of capitalism" (Crouch and Streeck 1997) among the modern core states since the seventeenth century in the sense that the institutional and cultural forms that emerged reflected earlier legacies with different material and cultural situations. And the same can be said of the trajectories of the individual countries of the Global South, where various forms of authoritarian regimes and economic and political institutions differed depending on legacies of the cultures that were incorporated into the Europe-centered system and of the colonial empires into which these regions were incorporated. The world history of capitalism should acknowledge these differences while still telling the story of the whole system.[4] First core states, and then noncore states, extended citizenship from wealthy land owners to men of no property, abolished slavery and serfdom, and eventually extended the vote to women (Ramirez et al. 1997; Schaeffer 2014). Authoritarian forms of government were mainly in the Global South, but the second-tier core states also experienced a resurgence of authoritarianism in the twentieth century. Both the Global North and Global South experienced "waves of democracy" (Markoff 2015b) and the eventual rise of parliamentary institutions and regime change by means of elections instead of military coups. These changes occurred first in the Global North and later in the Global South. Authoritarian regimes were more frequent in the Global South and still are (Almeida 2008b). But both the Global North and South experienced waves of parliamentary democracy and waves of recurrent authoritarianism in the nineteenth and twentieth centuries. Another wave of authoritarianism is happening now (Berberoglu 2020). The return of the "heroic" charismatic leaders is partly driven by parliamentary deadlocks that are more frequent and insoluble in periods of systemic crisis.

Centrist Liberalism adopted the social democratic welfare state as a response to both radical challenges to the property of elites from the Left, and popular fascist movements and regimes from the Right. Nationalism, the most powerful collective solidarity in the modern world-system, also evolved its institutional nature and meaning in response to challenges from both the Left and the Right.[5]

The Global Left evolved in response to the structural changes that occurred in the whole system and in response to changes in both Centrist Liberalism and the Global Right. Movements for popular sovereignty, to extend citizenship to men of no property, and eventually to women, and to abolish slavery and serfdom, melded with the rise of the labor movement and its variants—anarchism, socialism and communism. And this occurred in the context of the two great waves of movements for national sovereignty (decolonization) in the noncore (the first in the last quarter of the eighteenth century and the first quarter of the nineteenth and the second after World War II), including movements of indigenous peoples desiring autonomy (Hall and Fenelon 2009). Conservation movements differentiated into radical and centrist wings of the environmental movement.[6] Pacifism and Anti-War movements emerged to oppose warfare and violence. These Leftist movements formed an evolving constellation of what Arrighi, Hopkins, and Wallerstein (1989) called "the family of anti-systemic movements." These movements competed with one another as well as occasionally cooperating. A great wave of cooperation emerged as united and popular fronts in the 1930s in reaction against the rise of fascism. In the World Revolution of 1968 several new social movements (student rights, sexual orientation, and a now-global indigenous movement) joined the older ones, and the older movements (peace and anti-war, civil and human rights, third-wave feminism, labor, Third Worldism) coevolved with the newer ones.[7] The World Revolution of 1989 added a new appreciation for so-called bourgeois freedoms as well as cementing the notion that socialism and communism had failed. And more egalitarian movements emerged in the last decade of the twentieth century and the first decade of the twenty-first (slow food, food sovereignty, alternative media, the knowledge commons, global economic justice, and climate justice, see chapters 3 and 4). The constellation of movements changed in interaction with one another and in interaction with the version of Centrist Liberalism that emerged after 1968 (neoliberalism) (Harvey 2005). It also interacted with the evolving nature of the Global Right and all three factions of the political geoculture were conditioned by, and had consequences for, the global structural processes named above.

The Global Right emerged as a reaction against the French Revolution and the Napoleonic attempt to convert the core of the world-system into an empire. The British and the Austro-Hungarian Empires were the main supporters of the Concert of Europe, a precursor to the League of Nations that was intended to shore up the European traditional powers against movements for popular sovereignty.[8] The Austrians and the British fell out over how to handle the emergence of Italian nation-building and the Concert of Europe disintegrated. The conservatives in the nineteenth century continued their opposition to the rise of popular sovereignty and support for the power and prestige of religious hierarchies and aristocratic property and rule. The World Revolution of 1848 saw nascent workers movements allying with students and intellectuals to try to construct democratic nation-states based on popular sovereignty; but it was crushed by traditional authorities and their allies.[9] However, parliamentary democracy and labor unions and parties eventually spread across the core. Germany and Japan rose as challengers to the British hegemony. The booms and busts of capitalist industrialization produced losers as well as winners. The Left mobilized peasants, workers and the unemployed with ideas about socialism, proletarian internationalism, and anti-militarism. Passmore's (2012) account of the emergence of proto-fascist ideas in the nineteenth century is excellent.

The contemporary Global Right includes neoconservatives, conservative, and reactionary think tanks (Bob 2012) and media outlets, populist nationalists, anti-immigrant movements, neofascists, male supremacists, racial supremacists, climate change deniers, and reactionary religious fundamentalisms (jihadists, Hindu nationalists, and Christian identity groups). It also includes white identitarianism which combines elements of white supremacy, populist nationalism, religious fundamentalism, and neofascism. This broad constellation of contemporary counter-hegemonic far-right groups suggests comparisons with analogously diverse players in the World Revolution of 1917.

Logics of Accumulation: Coevolution of Capitalist and Tributary Modes

Fascism is a hybrid of capitalism and a version of the tributary mode of accumulation that has coevolved with capitalism for centuries and with the efforts to establish socialism since the late nineteenth and twentieth centuries. In order to develop a better understanding of twenty-first-century fascism, populist nationalism and authoritarian practices and regimes and possible twenty-first-century futures, it is helpful to understand this long-term process of coevolution among these different logics of accumulation.

The comparative evolutionary world-systems perspective sees human prehistory and history as having evolved from a kin-based mode of accumulation that regulated interaction by means of consensually held norms to tributary modes that added institutionalized and organized coercion (the law, state regulation, etc.) over the top of the kin-based institutions, to the capitalist mode in which accumulation became mainly based on the making of profits from commodity production and financial services.[10]

The tributary modes of accumulation have directly used state power (institutionalized coercion based on the law and its enforcement and specialized police and military organizations) to extract surplus product from domestic and distant populations through taxes, tribute, serfdom, and slavery. States and empires that mainly employed tributary accumulation have usually been controlled by military, priestly and land-owning elites whose wealth was mainly based on this form of accumulation. The tributary states and empires emerged during the Bronze Age out of kin-based chiefdoms. They frequently engaged in military competition (warfare) with one another for territorial conquest and tribute. In contrast, capitalism accumulated surplus value by making profits on the production of commodities and financial services while repurposing state institutions to support profit-making. Early capitalism grew in the interstices between tributary states and empires. Autonomous semiperipheral capitalist city-states under the control of merchants expanded regional markets and induced the production of surpluses for trade (Chase-Dunn and Hall 1997; Chase-Dunn, Anderson, Inoue, and Alvarez 2015). The growth of long-distance trade in the Iron Age encouraged the internal commercialization of tributary empires (Sanderson 1995).

Since the sixteenth-century capitalism has become the predominant logic of the Europe-centered (modern) world-system in a series of waves in which nation-states have increasingly come to be controlled by capitalists, and market forces and money have deepened and geographically expanded their influence. Warfare was increasingly employed as an adjunct to profit-making. The shift from classical imperialism (conquering ones' neighbors) to colonial imperialism (conquering distant sources of raw materials and trading nodes) accompanied the rising predominance of capitalist accumulation. The failure of the Habsburg Empire of the sixteenth century to establish a core-wide empire over the regions of emergent capitalism in Europe was an early instance of coevolution. Capitalist accumulation in the noncore (the periphery and the semiperiphery) was a mix of commodity production with forms of coerced labor (slavery and serfdom) that had been developed in the tributary empires. This mix has been called "peripheral capitalism" and is rightly seen as a necessary part of the global process of capitalist accumulation, but its overlap with the tributary mode is of interest to us because we are considering how the tributary mode has coevolved with capitalism.[11]

Tributary formations cycled back and forth between more and less centralized forms everywhere. The rise and fall of Rome as well as European feudalism and absolutism were examples of this. European capitalism was able to become predominant in part because the tributary states were weak in feudal Europe. The Dutch revolution moved the capitalist state from the semiperiphery to the core and facilitated the emergence of a world in which capitalist nation-states with colonial empires were the drivers, though old-style territorial empires survived until the nineteenth century.

Perry Anderson's (1974) *Lineages of the Absolutist State* argued that European "Absolutism" was an effort by feudal tributary kings to harness the surplus being generated by emerging capitalist enterprises. The Spanish and the Habsburgs were trying to reconstitute Europe into an updated tributary Roman empire but they failed and the system that emerged was a multicentric world-economy in which market forces were supplemented by a series of hegemonic core powers.

The tributary mode periodically regained purchase during periods of crisis in the modern world-system. The Habsburg Empire, the Napoleonic episode, and the German attempts in the twentieth century to establish a core-wide empire were instances of the tributary mode regaining importance and coevolving with capitalism. The efforts to build a socialist world-system in the twentieth century devolved into state communism in the Soviet Union and the People's Republic of China, resulting in regimes that combined elements of socialism with capitalism and elements of the tributary mode (Chirot 1977; Bahro 1980; Boswell and Chase-Dunn 2000). The fascist movements and regimes of the 1930s and 1940s were resurgences of the tributary mode in which institutionalized coercion in different forms was brought back as a major mobilizer of social labor.

Twentieth Century Fascism, Right-Wing Populism, and Neofascism

In the perspective of world history, fascism appears as a hybrid of capitalism and the tributary mode of production that emerged after economic institutions had evolved high levels of commodification. Twentieth-century fascism emerged from the second-tier core as reactionary populist-nationalist movements in a context of economic and political crises, which were then joined by big land owners and industrialists who sought to use them to counter the radical challenges coming from the left (Riley 2018).[12] Where fascist regimes came to state power, they tried to impose authoritarian control of the economy. Twentieth-century fascism was the most recent form that the tributary modes have taken. But the recent reemergence of right-wing populism, neofascism, and undemocratic authoritarian parties and regimes suggests that another wave of the tributary mode is emerging.

W. L. Goldfrank (1978), Michael Mann (2004), and Kevin Passmore (2012) have noted that twentieth-century fascism is difficult to define as a unitary phenomenon because it was not uniform but varied greatly by location and context and it evolved over time. It is particularly hard to define fascist ideology because the leaders were often extremely pragmatic and promiscuous in their choice of ideologies. One of the key features of twentieth-century fascism—hyper-nationalism—was constructed differently in dissimilar locations, which produced locally specific internal and external enemies.[13]

Definitions of twentieth-century fascism in the scholarly literature vary in width, choice of characteristics, and in emphasis on different characteristics. Some emphasize the nature of deeds (e.g., Paxton 2004) while others focus more on ideology. Michael Mann (2004: 13) strikes a useful balance between these: "I define fascism in terms of the key values, actions and power organizations of fascists. Most concisely, fascism is the pursuit of a transcendent and cleansing nation-statism through paramilitarism."

Both Mann's (2004) and Robert Paxton's (2004) excellent studies realize that this definition may not exactly fit somewhat similar phenomena that have emerged since 1945, but they are both willing to examine the nature of late twentieth- and early twenty-first-century neofascism to tease out the similarities and the differences. There were several "flavors" of fascism in the twentieth century and new flavors are emerging in the somewhat different context of the twenty-first century. One or two of the characteristic features may not be present in a contemporary right-wing movement but it may still be useful to designate it as neofascist.

In some countries, twentieth-century fascism was posed as secular national socialism and syndicalism, whereas in others it was formulated in religious terms. Italian and German fascisms were anticlerical. However, religion-based fascism did exist. A perceived "ideological crisis within the state" was tied to the rise of fascism in Turkey. Paxton (2004: 203) cites examples of religious fascism such as the "Falange Española, Belgian Rexism, the Finnish Lapua Movement, and the Romanian Legion of the Archangel Michael."[14]

Definitions of right-wing populism face similar issues. Snow and Bernatzky (2019), contemplating recent movements in the United States, settle on "an exclusionary form of identity politics" that supports a charismatic, messianic leader who claims that only he can solve the issues or ensure the promises of an emerging transpositional order. The "people" are seen as a pure collectivity that has been ignored by elites and the elites are seen as having favored another collectivity to the detriment of the "people." This definition, which is inspired by the contemporary polarized politics of the United States, overlaps to some extent with fascism as defined above, but there are also important differences, which we will discuss below.

Mann (2004) and Goldfrank (1978) distinguish true fascist regimes from authoritarian states that emerged from above in the first half of the twentieth century. They contend that real fascism bubbles up from below as a cult, a movement, a party, and then a regime. In the first half of the twentieth century, regimes emerged in Brazil, Argentina, and Japan that were authoritarian statist regimes but not true fascist regimes. They sometimes employed fascist imagery but were mainly movements from above, not from below.[15]

Populist right-wing movements emerged during and after the French Revolution when peasants from the Vendee rose to protect the king and the church from the Jacobins (Tilly 1964).[16] They were appreciated, supported, and manipulated by aristocratic and theological conservatives. But the fascist movements that emerged in the early decades of the twentieth century were different. They emerged under the leadership of non-elites, some of whom (like Mussolini) were former leftists. These right-wing movements emerged to confront the Left and in reaction to, and disgust with, the parliamentary stalemates and endless partisan jockeying for power of early electoral democracies (Mann 2004; Paxton 2004). The left was internationalist and supported immigrant workers and opposed militarism. Fascist populism glorified the nation, the state, political violence, advanced technology and militarism.

Fascist political leaders denigrated the importance of consistent political ideology and championed action and ideological pragmatism. They asserted the importance of a unified nation led by an authoritarian state that did not compromise with opponents. Fascists picked different internal and external scapegoats and used different forms of organization in the different countries in which they emerged. In the countries in which fascists were successful in building a mass movement and attaining state power, they did so by allying with powerful partners such as land owners and large enterprises; but traditional conservatives were chary about these alliances. Some of them sought to use the fascist movements against the left and to shore up their own privileges. But the tail sometimes wagged the dog. The support given by traditional elites was often motivated by fear of dispossession by the socialists and communists (often based on a paranoid misperception of the extent of the power of leftist forces, according to Michael Mann (2004: 21). Most traditional conservatives were nervous about the upstarts who led the fascist movements and wary of what populist fascist movements might do once they obtained state power. But the crises of global capitalism that produced World War I and the radical labor movement also produce twentieth-century fascism as a reaction against both Centrist Liberalism and the Global Left.

Some Marxists (e.g., Poulantzas 1974) have portrayed fascism as a repressive tool of industrial capital. Twentieth-century fascism emerged during a crisis of global capitalism, but it did not originate as an instrument of capital. It was a popular reactionary movement that sought to return to a fantasized purer earlier national solidarity and to oppose enemies who were thought to be threatening the purity of the nation. Some land owners and industrial capitalists supported the fascist movements and fascist regimes, but they were not its original source (Mann 2004: 20).

Fascist movements and regimes in the twentieth century were authoritarian attacks on democracy and the rule of law, but their hyper-nationalism eventually led to the further institutionalization of civic nationalism as an important form of modern collective solidarity. The scholarly analysis of nationalism discusses different forms: racial versus civic. Racial nationalism is understood as being based on genetic inheritance ("blood"), while civic nationalism is a collective identity organized around notions of consensual beliefs and commonly held legal and traditional norms and a shared history (Calhoun 1997). In practice, the populations of national societies often combine these types and their relative importance varies from country to country. It is civic nationalism that has become enshrined in the centrist geoculture, while the hyper-nationalism discussed as a feature of fascist movements often veers toward the racial side.

Fascist hyper-nationalism provoked a cosmopolitan reaction against extreme forms of nationalism that resulted in the civic form of nationalism becoming embedded in the Centrist Liberal part of the geoculture and the international institutions that emerged after World War II (the United Nations and the international financial institutions).

Comparing the Global Right with the Global Left

As we have said, the geoculture is an evolving constellation of contending ideologies of the right, center, and left. These assemblages interact with one another as well as reflecting, and having consequences for, the changing institutional structure of the modern world-system (Nagy 2017). One obvious difference between much of the Global Right and the Global Left is with respect to nationalism. The Global Left in the World Revolution of 1917 was explicitly internationalist, though it did not deny the possibility of progressive nationalism. Most socialists, communists, and anarchists believed in proletarian internationalism and condemned conservative forms of nationalism as false consciousness that was promoted by capitalists to undermine the class struggle and to get workers to go to war against one another. This was an important instance of secular global humanism and cosmopolitanism, though it was mainly understood as international class solidarity. Proletarian internationalism came to grief when the German government tricked the German socialists into voting for war credits at the outbreak of World War I, thus abrogating an agreement among the national parties of the Second International to not go to war and kill each other at the behest of their national capitalists. It was this development that sealed Vladimir Lenin's disgust with the labor movements of the core and provoked his turn to the "Third Worldism" of the Third International (Claudin 1975). As we have discussed in chapter 4, internationalism, transnational humanism, and Global Southism (formerly Third Worldism) have continued to be important political commitments of the New Global Left in the World Revolution of 20xx (Steger, Goodman, and Wilson 2013; Carroll 2016).

We have already said that nationalism is the most important collective identity in the modern world-system (Calhoun 1997). With the rise of popular sovereignty in the world revolutions of 1789 and 1848 states became legitimated as agencies for protecting and developing the interests of "the people" and the people came to be understood in terms of nations. While the nation-state became the normative ideal, many actual states included people of different ethnic groups and cultures. The idea of ethnic minorities emerged in contrast to national majorities. Progressives have most often noted that there have been both reactionary and progressive versions of nationalism. As Tom Nairn (1998) contended, political nationalism is Janus-faced. National liberation struggles in the Global South and progressive movements in the Global North both contended that nationalism could be harnessed as a progressive form. But internationalism and a critique of nationalism, also emerged from the Left in the nineteenth century and it continues to be an important feature of the

Global Left in the twenty-first century. Historical comparative research has found waves of nationalist movements that correspond with periods of disorder and crisis in the world-system (Karatasli 2018). Humanism, internationalism, and cosmopolitanism emerged in response to the further institutionalization of nationalism over the past two centuries. Radicals in the labor movement were committed internationalists, but they did not discourage their adepts from organizing national unions and political parties. Rather they sought to provide support from abroad to progressive national movements.

Fascist movements before and after World War I attacked the workers movements and socialist parties both because the fascists opposed class struggle in favor of organic nationalism and because they opposed the internationalism and pacifism of the Left (Paxton 2004). Attacking peasant unions and labor unions also gained the fascists the support of land owners and some large capitalists, and this became an important source of powerful elite support and finance for those fascist sects that were able to move on to become mass movements and to take over national regimes; but hyper-nationalism was also an obstacle to transnational and international cooperation and organization. The fascists did try to organize a fascist international during the late 1920s and the 1930s (Laqueur and Mosse 1966), but their strong commitment to the myths of nationalism stood in the way (Paxton 2004: 20, fig. 83).[17] This was, and still may be, an important difference as well as a conflict between the Global Right and the Global Left.[18] Its current main manifestation is about the treatment of migrants.[19]

Dani Rodrik (2018) contends that two kinds of populism have arisen to contest the neoliberal globalization project. During the Latin American debt crisis of the 1980s and the 1990s the structural adjustment policies of the International Monetary Fund required draconian austerity and privatization (as discussed in chapter 2). These policies were supported by neoliberal national politicians who attacked the labor unions and parties of formal sector workers, but this produced a populist reaction in many countries in which progressive politicians and leftist opposition parties were able to win elections by campaigning against these policies and by mobilizing the rural and urban popular classes (Walton and Seddon 1994; Davis 2006; Almeida 2010b). This phenomenon has been called the "Pink Tide" and the relationship between leftist parties and movements has been called "social movement partyism" (Almeida 2010a). Regimes based on left-populism emerged in most of the Latin American countries (Levitsky and Roberts 2011), and Rodrik rightly sees this as a reaction against the neoliberal globalization project. Christopher Chase-Dunn, Alessandro Morosin, and Alexis Alvarez (2015) found that semiperipheral countries experienced the Pink Tide before peripheral ones did.[20] They contend that the Pink Tide phenomenon emerged in Latin America and not in other regions of the Global South because Latin America has proportionately more semiperipheral countries than Asia or Africa. And the more draconian structural adjustment polices of the 1990s were tried first in Latin America (see figure 2.4 structural adjustment map in chapter 2).

Right-wing populism emerged, and is still emerging, in countries of the Global North in which neoliberal globalization produced deindustrialization and many workers lost their well-paid jobs in manufacturing (Van Dyke and Soule 2002; McVeigh and Estep 2019). In Europe and the United States, workers in older industries who had been decimated by capital flight to low-wage countries, were mobilized by right-wing politicians who blamed immigrants and liberals for what had happened during the neoliberal globalization project (Berezin 2009; Moghadam 2020a). This occurred in contexts in which it was easier for politicians to blame immigrants and minorities than to point the finger at the big winners of the neoliberal globalization project. And some of the big winners provided support for the politics of hyper-nationalism, xenophobia, racism, status anxiety and sexism that are the working muscles of right-wing populism and neofascism (Alvarez 2019).

Right-wing populist politicians have exploited cleavages along racial, ethnic, and cultural lines, rallying individuals against foreigners and minorities (Bobo 2017). Left-wing populist movements, on the other hand, have tended to mobilize support based on economic cleavages. They have targeted large corporations and finance capital as responsible for economic dislocations and collapses. The ease of mobilization around these cleavages depends on the salience of the issues for constituents. Individuals manipulated by rightist demagogue politicians and racist social media platforms who feel their jobs and public services have been threatened by immigrants and minorities are easier to mobilize along ethno-national and cultural cleavages.

The cleavages Rodrik describes are easier to exploit in areas impacted most by massive migrations and economic relocation and divestment. Global warming has led to declining agricultural output, declining peripheral state revenues for local constituencies, natural disasters, and the collapse of failed peripheral states leading desperate people to flee local conflicts and poverty by seeking refuge in the Global North. In the North, rapid automation has led to unemployment among both skilled and unskilled workers (Grimes 1999). Right-wing politicians have been able to prey on the fears of economically insecure individuals in the Global North, who see waves of migrants as threats to their social and economic way of life (McVeigh and Estep 2019).

Reactionary Islam and the Global Right

Reactionary Islam is a counter-hegemonic movement but not a progressive one (Moghadam 2009).[21] In drawing the attention of the neoliberal powers that be after 9/11, the jihadists provided a bit of space for the Latin American Pink Tide. But neofascism and right-wing populism are unlikely to find allies among the jihadists. Indeed, anti-Islamic rhetoric has mainly come from the alt-right's racist and anti-immigrant forms of nationalism.

Immanuel Wallerstein (2017) explained the rise of Islamic fundamentalism as a reaction to the economic downturn of the 1970s that signaled the beginning of another Kondratieff Wave B-phase, the collapse of the Soviet Union, and the failure of the movements of the World Revolution of 1917 to produce a more humane and sustainable global society, and so their loss of popular support. Wallerstein contended that religious identities had been becoming less and less important in world politics from the Protestant Reformation until 1970.[22] He also points out the ambiguous relationship that religious fundamentalism has had with states. God's law is allegedly higher than state law, but the fundamentalists try to take state power to impose god's law.

Religious fundamentalism since the 1970s has also been a reaction against the world revolution of 1968, as was the rise of neoliberalism. But neoliberals champion the further commodification of traditional functions formerly handled by families or communities, while fundamentalists oppose commodification. And many fundamentalists oppose gains for gender equality, racial equality, and nontraditional sexual orientations that are perceived as disempowering traditional religious authority. Religion plays an important role in contemporary right-wing social movements as a force for mobilization, cohesion-building, and the effort to restore the authority of patriarchal families and tradition religious leaders. Religions provide frames and imagined golden ages that are used by many right-wing movements to build their ideologies (Denemark 2008). One reason why some religious fundamentalisms became counter-hegemonic after 1970, but were not very important in the World Revolution of 1917, is that fundamentalism became a functional counter-hegemonic substitute for revolutionary Marxism and related Leftist ideologies that had been the basis of the Global Left in the twentieth century (Grimes 2003). Wallerstein contended that one of the causes of the rise of politically reactionary religious fundamentalism was the perceived failure of secular counter-hegemonic movements, especially communism. It is as if there is an ideological menu embedded in the geoculture from which individuals and groups can appropriate frames. When one appears to have been discredited, others are selected.

The Current World Revolution

The global political, economic, and demographic situation has evolved in ways that challenge many of the assumptions that were made during the rise of the Global Justice movement and that will require adjustments of the analyses, strategies, and tactics of progressive social movements. The Arab Spring, the Latin American Pink Tide, the Indignados in Spain, the Occupy movement, the rise of progressive social movement–based parties in Spain (Podemos) and in Greece (Syriza), and the spike in mass protests in 2011 and 2012 (see figure 2.1) inspired some activists to label the contemporary world revolution that emerged in the last decade of the twentieth century "the World Revolution of 2011." The electoral victory of the progressive Syriza Party in Greece in 2015 led to a debacle in which the effort of the new regime to renegotiate the terms of the national debt was crushed by the European banks. The banks doubled down on austerity, threatening to bankrupt the pensioners of Greece unless the Syriza government agreed to new structural adjustment austerity policies, which it did. This was a case in which another world may have been possible, but it did not happen. This outcome was a slam on the other New Leftist social movement parties in Italy and Spain as well as on the Global Justice movement. The new progressive parties are trying to democratize the European Union so that it represents more than just the banks and the large corporations (DiEM25 2015). This effort to increase the legitimacy and democracy of the European Union is happening in the context of Brexit and nationalist reactions to the European Union. This may be an opening for the European Left, especially Green Parties.[23]

The huge spike in global protests in 2011–2012 was followed by a lull and then a renewed intensification of citizen revolts from 2015–2016 (Youngs 2017), and again in late 2019 with the outbreak of massive austerity protests in several world regions against economic inequality. The Black Lives Matter movement, the Dakota Access Pipeline protest, the Me-Too movement, the anti-Trump Women's Marches, largest strike in world history in India, the Antifa rising against neofascism, the Climate Justice movement, and the somewhat ambiguous rise of popular protests that have been sparked by rising fuel costs (the Yellow Vests in France and many similar demonstrations elsewhere such as in Sudan, Mexico, Guinea, Haiti, Iran, and Chile) suggest that the World Revolution of 20xx is not over. But the setbacks and the rise of populist nationalism, anti-immigrant movements, and neofascist and authoritarian regimes that play to these movements, along with defeats of progressive movements and the demise of the Pink Tide in Latin America, require a reassessment of the context and strategies of the progressive movement of movements (Bond 2019). Such reassessments will be much more pressing given the economic slowdown and mass unemployment triggered by the novel coronavirus.

The mainly tragic outcomes of the Arab Spring and the decline of the Pink Tide progressive populist regimes in Latin America have been distressing for the Global Left. The Social Forum process and progressive mass demonstrations in favor of democracy were late in coming to the Middle East and North Africa, but they eventually did arrive in the form of the Arab Spring. The demonstrations were mainly rebellions of progressive students using social media to mobilize mass protests against old authoritarian regimes (Korotayev and Zinkina 2011). The outcome in Tunisia, where the sequence of protests started, has been somewhat encouraging. But the outcomes in Egypt, Syria, Turkey, Iran, and Bahrain were disasters (Moghadam 2017; 2018).[24] Some of these popular democratic movements were defeated by repression from the old regimes and some by Islamist movements that were better organized (Egypt) and by outside intervention (Syria). In Syria, the movement was able to organize an armed struggle, but this was defeated by the old regime with Russian help. Extremist Muslim fundamentalists took over the fight from progressive students. The Syrian civil war produced a huge wave of refugees that combined with economic migrants from Africa to cross the Mediterranean Sea to Europe. This added fuel to the already existing populist nationalist movements and political parties in Europe, propelling electoral victories inspired by xenophobic and racist anti-immigrant sentiment. In Iran, the green movement was repressed (Harris 2012; Honari 2019). In Turkey, Erdoğan has prevailed, repressing the popular movement and continuing to fight the Kurds. All these developments, except Tunisia, have been setbacks for the Global Left.

The Pink Tide in Latin America emerged in the last decade of the twentieth century and the first decade of the twenty-first century when progressive populist politicians and oppositional parties were able to mobilize the urban poor to support expanded welfare states based on the export of raw materials (Almeida 2010a and chapter 4 of this volume). These regimes did not much challenge the global power structure and did not try to dispossess their domestic elites, but they did provide services and encouragement to traditionally marginal groups and support for the Social Forum process despite its formal refusal to allow participation by elected political authorities. These regimes emerged in reaction to the crackdowns on state subsidies and labor unions that were supported by domestic neoliberals and by the early structural adjustment programs imposed by the International Monetary Fund (discussed above and in chapter 2). This is a new twist on the idea of the resource curse. Populist welfare states that rely on the export of raw materials can fund programs for the poor without dispossessing their national elites, but they remain dependent on the world market and the global financial system, so they are not able to survive when commodity prices plunge. This is the latest incarnation of the boom and bust cycle experienced by the periphery and semiperiphery for hundreds of years.

The replacement of most of the Pink Tide progressive regimes in Latin America by reinvented local neoliberals and right-wing populist politicians has largely been a consequence of falling prices for agricultural and mineral exports because Chinese demand has slackened. The social programs of the leftist populist movements were dependent on their ability to tax and redistribute returns from these exports (Huber and Stephens 2012). But both the rise of the Pink Tide and its demise may be an improved new normal for Latin America because almost all earlier transitions involved military coups and extensive violent repression, whereas the rise of the Pink Tide and most of the more recent rightward regime transitions have been relatively peaceful and have not involved takeovers by the military or much violent repression (with the exceptions of Honduras, Nicaragua, and Bolivia). The legal shenanigans that led to the electoral demise of the Workers' Party government in Brazil were not pretty (and the threat of a military coup has continued to play a role in politics) but, at least so far, the rightward shift has been less violent than earlier regime transitions. The interventionism in Venezuela to take down the remains of the Chávez Pink Tide regime is an unsavory reoccurrence of US imperialism toward its Latin American "back yard" (Bean 2017; Ciccariello-Maher 2017). The outcome may involve a civil war or just a negotiated regime change. Not everything has changed in North/South relations. But relatively stable parliamentary democracy may have finally arrived in most of Latin America. This is not utopia, but it is progress. Leftists can contend for power in the next round.

The Pink Tide inspired some on the left to hope that the rise of the BRICS might provide an opening for the Global Left and a powerful force that could confront the neoliberal globalization project (Chase-Dunn 2013).[25] But as the BRICS regimes faced economic pressures due to their dependence on raw material exports, they increasingly adopted aspects of neoliberalism in both Latin America and Africa (Bond 2013). These political retreats inspired protest movements and charges of corruption and the eventual rise of right-wing populist politicians who promised to drain the swamp.

The continuing rise of right-wing populist and neofascist movements and their electoral victories in both the Global North and the Global South have added a new note that is reminiscent of the rise of fascism during the World Revolution of 1917. This raises the issue of the relationships between movements and counter-movements (Bob 2012; Nagy 2017). International nongovernmental organizations (INGOs) emerged as important players in world politics in the last half of the nineteenth century (Murphy 1994). As Clifford Bob (2012) points out, most studies of INGOs and global governance focus on progressive advocates of human rights, etc. The world of INGOs is extremely contentious, with conservative INGOS competing with progressive and centrist ones to shape the policies of states and of international governmental organizations.

The glorification of strong leaders in the right-wing populist and neofascist movements was also a characteristic of twentieth-century fascism. But charismatic leaders have also been important for progressive movements in the past and probably will be in the future despite the "leaderless" ideology of the horizontalists (e.g., the rise of López Obrador and the Morena Party in Mexico). The polarization of politics provoked by the rise of neofascist and right-wing populist politicians is increasingly providing an opening in the mass media for discussion of the ideas of progressive movements and political leaders.[26]

Horizontalism and the appreciation of spontaneity have a long history in the Global Left. Anarchists have long disparaged official hierarchies and hierarchical political organizations of progressives. Some leftists have always glorified the spontaneity of mass demonstrations and unorganized collective behavior (Piven and Cloward 1977), while Rosa Luxemburg posited a dialectical relationship between spontaneity and organization (Hudis and Anderson 2004). Direct participatory democracy was an important component of the New Left's critique of the Old Left in the World Revolution of 1968. But the recent disappointments of the Global Justice movement and the rise of right-wing populism and neofascism have provoked a new effort to combine horizontalism and spontaneity with an organizational structure that can facilitate cooperation and mobilize resistance and contention with the powers that be in world politics.[27]

There has always been a tension within the Global Left regarding anti-globalization versus the idea of alternative progressive forms of globalization. Samir Amin (1990) and Waldo Bello (2002) have been persuasive progressive advocates of deglobalization and delinking of the Global South from the Global North. Amin supported the idea of progressive national projects that would empower workers and farmers and carry out collective development projects that would serve the people. The alter-globalization project has been articulated by Geoffrey Pleyers (2010). This globalist approach focusses on global justice and global inequalities and supports transnational social movements and engagement with international organizations.

While the United States has exercised a de facto global military empire since the demise of the Soviet Union, US economic hegemony has been in decline since the 1970s. The proportion of the global domestic product that is accounted for by the US economy has declined since its peak in 1945 (Chase-Dunn, Kwon, Laurence, and Inoue 2011). In the 1970s, German and Japanese manufacturing caught up with US manufacturing, leading to global overcapacity and the expansion of finance capital. The rise of economic competitors from the BRICS has produced a multipolar global economy that will eventually be followed by a multipolar distribution of military power, because the current concentration of military capability under the control of the United States is very expensive, and has mainly been made possible by the survival of advantages accrued during the long period of US economic hegemony (Chase-Dunn and Inoue 2017). The coming of multipolarity may be an opportunity for countries in the Global South but it could also be a very dangerous situation if rivalry for global domination among powerful states leads to interstate warfare.

Neofascist Movements

Most neofascist movements do not simply regurgitate the rhetoric of the early twentieth-century fascist movements. They have been shaped by the contemporary socio-political-economic structural context and the ideas that have become taken-for-granted in the twenty-first-century geoculture (Paxton 2004; Alvarez 2019). Neofascist movements have not (yet) been as violent, and nor have they glorified violence as much as their twentieth-century predecessors did.[28] They are generally covert and adaptive, attempting to capture public spaces. In the old days that meant streets, villages and newspapers. In recent decades, it has meant inexpensive rural television and radio venues, the internet, and online social networks (Caren, Jowers, and Gaby 2012).

Jerry Harris et al. (2017) claim that the leadership of neofascist movements relies on misdirection, and on their supporters' comfort with alternative facts so that they can survive in a globalized economy while pushing isolationist and racist agendas. As with the New Global Left, they are a reaction to the neoliberal globalization project, but instead of proposing an alternative form of democratic and multicultural globalization they propose reactive nationalism, he-tooism, xenophobia, protectionism, and "making × great again."

Another difference between the earlier and more recent versions of fascism is the attitude toward the national state. Most of the earlier versions glorified the state as an instrument of the purified nation. The realities of state control were more complicated in both Italy and Germany, but the ideal of "statism" was an important fascist value. Contemporary neofascist movements do not glorify the state. They favor more authoritarian and interventionist state actions, but they do not glorify the state as such. This difference is one reason why some scholars prefer the term "populist nationalism" over "neofascism." Another important difference is about military expansionism. Glorification of military expansionism was an important part of Italian and German fascism and Japanese authoritarianism. Argentine efforts to reclaim the Malvinas, Russian military expansionism in Ukraine and the attempt to establish an Islamist caliphate in Syria and Iraq are interesting counter-examples but, compared with the glorification of conquest that was espoused and carried out by Italy, Japan, and Germany in the twentieth century, there is less real militarist expansion now. Neofascist movements or parties have not yet endorsed military expansionism, at least so far. The decolonization of the whole periphery and the establishment of international organizations such as the United Nations that oppose conquests and support the sovereignty of member states has largely delegitimized formal colonialism and conquest. Clientelism and covert interventions continue to be the main modes of exercising power in geopolitics. It is likely that neofascist regimes would not hesitate to employ covert interventions and saber-rattling, but a return to large-scale military conquests seems unlikely (Kumral 2015).

Though many of the earlier fascist movements embraced syndicalism and were anti-capitalist in their early phases, most neofascists and right-wing populists now strongly support capitalism and oppose state intervention in the economy (Hochschild 2016; Skocpol and Williamson 2016; but see Minkowitz (2017). This appears to be part of a continuing reaction against the welfare state that was pioneered by the rise of neoliberalism in 1970s and 1980s and continues to be an important theme in right-wing populist and neofascist movements.

Contemporary Christian and white identitarianism are recent forms of reactionary populism that seem to reflect an adaptation to the multiculturalism of Centrist Liberalism. White identity movements assert that people of European descent have an essential culture that needs to be defended against other cultures. Identitarianism is part of the global alt-right and has generated multiple versions, new religions, political parties, and legitimating ideologies such as the "white genocide" conspiracy theory. Generational identitarianists in France and Germany organize communal living for young people and assault foreigners, especially immigrants from the Islamic majority states (Bennhold 2018), while hate crimes continue at heightened levels in the United States.

The Evolution of the Global Right

The rise of twenty-first-century fascism has been caused by the crisis of the neoliberal globalization project. An "interlocking set of new enemies" is seen as tearing at the status quo, including "globalization, foreigners, multiculturalism, environmental regulation, high taxes, and the incompetent politicians" (Paxton 2004: 181). The neoliberal globalization project has led to a transformation of the world economy, thereby providing "a new fertile terrain for far-right mobilizations" (Saull 2015: 631). The fragmented, insecure precariat no longer gathers in membership-based collective organizations (Standing 2011). A globalized economy provides opportunities to blame immigrant labor, finance capital, foreign investment, labor-outsourcing, and ineffective politicians for local economic dislocations (Saull 2015).[29]

Enzo Traverso (2017) discusses the evolution of racism in the comparison of twentieth- and twenty-first-century fascisms. Anti-Semitism was the most important scapegoat in the twentieth century. While it is still around, Traverso contends that it has been largely replaced by Islamophobia as the central scapegoat trope of what he calls "post-fascism." This is an important difference because in the twentieth century Jewishness was associated with the Left, whereas in the twenty-first-century reactionary

Islam is not associated with the Left. Indeed, we have argued that it is part of the New Global Right (and see Moghadam 2009) even though the right-wing populists and neofascists are using Islam as a scapegoat.

Classical fascist rhetoric claimed to transcend class struggle (Mann 2004). But a divide now exists between those qualified for open sector, internationally competitive jobs and those stuck in sectors that are unable to compete in the neoliberal global marketplace (Kriesi et al. 2006). Globally-minded liberals and progressives have become the enemies of locally-focused traditionalists (Hochschild 2016). Individuals can only vote in their local and national elections. The European Parliament is a partial exception, but international organizations such the United Nations are lacking in their institutional ability to directly represent citizens (Monbiot 2003).

New Right movements seek to demonize characteristics of Centrist Liberalism such as materialism, individualism, the universality of human rights, egalitarianism and multiculturalism. These movements claim to restore the primacy and purity of ethno-national groups, now threatened by globalization and immigration. The immigration issue is not likely to go away soon. Population continues to grow rapidly in eight countries in the Global South in which prospects for economic development and job growth are very bleak and many other countries have large numbers of people who are of little use to the accumulators of capital. The peak migration wave of 2015 has receded, but economic forces, political disruptions, global warming, and civil wars are likely to continue to make South/North migration a contentious issue for the next several decades (Mason 2015).

Contemporary right-wing populist politicians who are trying to build a mass movement usually distance themselves from the political violence of their most ardent supporters.[30] The neofascist fringe groups use forceful confrontation as a tactic and this is reminiscent of the fascists of the twentieth century.[31] There is fear of growing securitization and more centralized and coordinated forms of surveillance of troublemakers. William I. Robinson (2013, 2014) theorizes the rise of a coordinated transnational capitalist state that is increasing its capacity for global repressive coordination. This may be the future but, so far, the geopolitical competition among contending nation-states seems to reflect a growing contentious multipolar geopolitical structure rather than a centralized and coordinated global structure of repressive power. Some neofascist organized actions have displayed considerable political savvy in framing and coordinating their street tactics.[32]

The energy that some neofascist groups have shown and the rapid spread of authoritarian rightist populism and neofascist movements across the globe portend a shift in the zeitgeist that progressive forces need to understand and mobilize against (Berberoglu 2020). The pre-COVID-19 low unemployment rate, the revival of the real estate market, and the Trump-induced stock market bump slowed these movements down. But the economic slowdown caused by the COVID-19 pandemic, as well as the electoral contests being held during a time of great polarization, will increase the level of frustration and support for these neofascist sects and movements.

The Future of the Global Right

The observation that efforts to revive the tributary mode of production reoccur in waves and that the nature of the institutional and ideological forms of these reoccurrences coevolve with Centrist Liberalism the Global Left has implications for the future. The tributary mode first became predominant in a regional world-system in Bronze Age Mesopotamia around 2500 BCE. Capitalism first became predominant in a regional world-system around 1500 CE. So, the predominance of the tributary mode lasted four thousand years whereas capitalism has only been predominant for about five hundred years. Backsliding could occur, but the stable reestablishment of tributary national states or a tributary global state would seem unlikely because capitalism and the forms of political legitimation that have emerged to contest it (democracy and the rule of law) are superior regarding their effective legitimation of authority relative to those of the tributary modes. This said, a temporary reemergence of undemocratic repressive regimes is already happening and may become a stronger trend if very unstable economic and security conditions emerge, which seems likely.

The institutions that were the backbone of the tributary mode were states, empires, the law, courts, prisons, and the military. The ideological bases of these were hierarchical religions but, since the emergence of Iron

Age world religions, these also produced a larger moral order in which both the emperor and the masses were members of the same moral community. The rise of capitalism caused these institutions to be repurposed to support profit-making but did not eliminate them. The evolution of political ideology from legitimation from above to legitimation from below (from the divine right of kings to popular sovereignty and the welfare of the people) and the rise and institutionalization of human rights place important limits on the forms and actions that evolved instances of the tributary mode can take. But twentieth century fascism demonstrated that modern forms of the tributary mode are indeed possible using secular ideologies such as fascism and national socialism.

Literature and the cinema are full of semi-believable fantasies about future repressive regimes. It is not hard to imagine powerful new forms of authoritarianism that could emerge at the global level or within single states. Indeed, newly authoritarian regimes have already emerged in the semiperiphery and the periphery in the last decade (Berberoglu 2020). Recently developed technologies of surveillance are already being employed in some countries to try to control the behavior of the masses (e.g., the People's Republic of China). Fears of global authoritarianism combine and interact with the other crises that are emerging in the twenty-first century. Huge global inequalities, anthropomorphic environmental degradation and global warming, a human population that continues to grow despite the demographic revolution and the aging demographic structure, a huge balloon of fictive wealth propped up by global banks and financial institutions, and the rise of right-wing populism and neofascism—these developments are daunting and cause many to see the human future as doomed. Indeed, a perfect storm in which financial crises, global warming, and interstate warfare could knock us way back, or even eliminate all life on Earth (Kuecker 2007).

But the Global Left could also mobilize citizens against neofascism and in favor of an organized collective rationality that would allow our species to deal with the problems of the twenty-first century (Standing 2014). Just as we can imagine new draconian political regimes, we can also imagine a federal global democracy in which technology and authority are used to meet the needs of people and to organize a sustainable world economy that will not destroy the ecological basis of human life. This would probably not be an ideal utopia and new problems would undoubtedly emerge and be contentious in world politics in the twenty-second century. What needs to be done is to get through the first half of the twenty-first century without making the big mistakes that were made in the first half of the twentieth century (Hobsbawm 1994). Malthusian corrections have reoccurred periodically in human sociocultural evolution, and the first half of the twentieth century was a global catastrophe that we must not repeat. Indeed, a twenty-first-century version could be worse because of the emergence of more deadly military technologies. The interregnum that will follow the further hegemonic decline of the United States will be a trial in which all the institutions of global governance will be tested. The outcome of the struggle might be a collapse or a new flavor of global fascism. But these would likely be temporary, and eventually humanity will move on to either a new systemic logic of human interaction (democratic socialism) or another systemic cycle of capitalist accumulation (Arrighi 2009). The Global Left will fight for a real systemic transformation and effective contestation will be necessary to motivate serious change. But a somewhat more humane and sustainable form of capitalism (e.g., Kenworthy 2019) would make a subsequent effort at qualitative transformation possible. A reemergence of tributary national regimes or even a tributary global state are also possible during the coming time of troubles in the first half of the twenty-first century. But these too would likely be temporary if the struggle they provoke does not lead to a fatal human catastrophe.

What can we conclude about the ways in which social movements and collective action have influenced sociocultural evolution and the path of world history over the past several centuries, and how will these phenomena have consequences for what happens during the rest of the twenty-first century?

SIX

The Future of Global Change and Social Movements

··

Lessons Learned from Premodern Collective Action

We have presented a broad and ambitious portrayal of global level dynamics and their influence on collective action as well as how those actions shape global structures. Modern social movements since the nineteenth century are often assumed by sociologists to have a set of unique features that legitimate dismissing the important roles played by social movements and collective behavior in premodern historical social change (Soule and Soule 2010). It is often assumed that repertoires such as the strike, the demonstration, and the boycott were modern inventions despite the fact that historians have documented such collective action going back centuries (MacMillan 1963; White 1995). We trace collective action back to premodern small-scale polities and show that social movements have been important causes of sociocultural evolution for thousands of years. These premodern movements are not simply important to empirically establish (although this is a substantial scientific advance in archaeology, anthropology, and historical sociology), but to show repeated patterns in world history over the long term as well as some of the core elements of the mobilization process that have survived into the twenty-first century.

Central features driving collective action and social movements for millennia include competition for resources and power, elite sponsorship, charisma, emotional energy, religious beliefs, disruption, and threats. Just as in previous centuries, elite-induced social movements continue to exercise influence in the new millennium. Wealthy conservative donors, such as the Koch brothers in the United States, have funded the Tea Party (Almeida and Van Dyke 2014; Skocpol and Hertel-Fernandez 2016) and paid gun-toting demonstrators demanding that public health regulations intended to slow the spread of COVID-19 be rescinded. President Donald Trump denounces international free trade agreements and targets immigrants to gin up sustained emotional support (Tarrow 2018). The power of charisma also serves in contemporary struggles against neoliberal globalization. Hugo Chávez Frías rose to electoral triumph with his Fifth Republic movement in Venezuela in the late 1990s by giving charismatic speeches that proclaimed neoliberalism to be a savage form of capitalism. Bolivian President Evo Morales used similar mobilization appeals in his Movement Toward Socialism (MAS) throughout the early 2000s. Populist right-wing demagogues throughout Europe and North America are mobilizing local populations with emotion-laden xenophobic and racist language against international migration from the Global South with increasing and alarming electoral success (Gest 2016). Scholars of global change and collective action, and students of social movements in general, need to give more recognition to the power of charisma and emotional appeals in catalyzing mobilization (Jasper 2018).

As we contended in the preceding chapters, religion and spirituality also drive global change. In the late twentieth century, the distortions of the global pattern of state-led development that favored the cities over the countryside provided the Catholic Church an opportunity to reinvent itself with the Second Vatican Council Reforms and the "option for the poor" professed by a Liberation Theology, which launched massive peasant rebellion throughout Latin America and beyond. In chapter 5, we also argued that reactionary Islam emerged as a major force against neoliberal globalization that grew in the context of the vacuum of counter-hegemonic ideologies after the leftist liberation movements of the World Revolution of 1917 failed to deliver a more egalitarian and just world order.

Consistent throughout this work, and emphasized in premodern movements, is the role played by global disruptions and threats in sparking the emergence of collective action. The disruption of everyday life, especially economic arrangements, makes alternative forms of action desirable by communities (Snow et al. 1998). Early forms of colonialism (especially settler colonialism) and empires created massive dislocations for local populations, motivating indigenous campaigns of resistance as well as dissident rituals. Global dislocations and threats continued in the 2010s, with climate change, drought, and mega-development and resource extraction projects tied to sustaining capital accumulation into the 2020s. The broad processes we observed in premodern societies in which elementary forms of collective action and social movements led by spiritual leaders, charisma, religion, and disruption continue to be relevant in the current century.

Anti-neoliberal National and Local Revolts

In chapter 2 we focused on neoliberalism as a global trend that spurred national and local revolts around the world with an emphasis on the Global South. We demonstrated the origins of the neoliberal transition in the erosion of Fordism and the global debt crisis of the 1980s. Special attention was given to the preceding period of state-led development and the extension of social citizenship rights. The state-led development period created both the organizational infrastructure and the motivations for launching rebellions decades later against neoliberalism, austerity, and structural adjustment. In particular, the expansion of the state infrastructure, especially transportation, mass public education, and public health facilitated collective mobilization. The erosion of social citizenship rights (welfare state retrenchment and austerity) provided the motivational impetus to initiate defensive campaigns to attempt to protect economic and social gains won in the previous decades.

We outlined the major anti-neoliberal revolts back to the 1970s with general strikes in Europe and early anti-neoliberal rebellions in the Global South such as in Egypt and Peru (Walton and Seddon 1994). Original protest event data highlighted massive outbreaks of anti-neoliberal protest campaigns spreading throughout the globe between the 1970s and 2010s. We demonstrated that structural adjustment became a permanent feature of the Global South in the 1980s and 1990s with most countries under several years of IMF and World Bank conditionality between 1980 and 2008. Chapter 2 also chronicled the continuing national revolts against neoliberalism through 2019 (as observed in Honduras, Ecuador, Chile, Lebanon, and Iraq). Several recent anti-neoliberal revolts center on price hikes of fuel and food as witnessed in the *Gasolinazo* in Mexico in 2017, and the fuel and price hike protests in France, Haiti, Sudan, Chile, Iran, and Guinea in 2018 and 2019.

Another critical feature of anti-neoliberal rebellion centers on local-level resistance. Global neoliberal forces work their way down to the village and community level. While a plethora of case studies exist of community resistance to neoliberal measures of subsidy cuts, free trade, incursion of foreign mining operations, and other austerity measures, few studies examine local-level variation in mobilized opposition to economic liberalization policies. We contributed to this literature by offering the recent example of Costa Rica and its IMF-counseled Fiscal Reform Package in late 2018. This neoliberal austerity package resulted in the largest civic and labor strikes in modern Costa Rican history. We showed that the key elements of resistance at the local level were found in segments of the state-led development sector (public schools and highways) as well as in strategic mobilizing experience via past community participation in mobilizing against the Central American Free Trade Agreement (CAFTA). These sectors have been found to be key contributors in explaining the different levels of resistance across communities in other anti-neoliberal protest campaigns (Almeida 2012, 2014) and should be given consideration in theorizing progressive coalitions into the future that confront the threats of global economic change and the decline in social citizenship rights.

Global Warming and Climate Justice

Chapter 3 introduced perhaps the largest global threat of all, climate change and global warming. The chapter largely focused on the development of a transnational movement to address and turn back increasing carbon emissions. We demonstrated that the Climate Justice movement is rooted in the Global Economic Justice and Anti-War movements of the early 2000s. The development of a worldwide internet communications technology infrastructure along with the emerging harms associated with global neoliberalism and free trade launched the Global Justice movement in the late 1990s and sustained it through the early 2000s. Most importantly, the Global Justice movement established a coordinating template—a repertoire of contention—that became institutionalized in the New Global Left. The coordinating structure came to be known as a Global Day of Action. It centered on holding massive demonstrations at the site (usually a major world city) of an elite global trade conference while simultaneously holding solidarity protest events around the world (Almeida and Lichbach 2003). The 1999 Seattle World Trade Organization Ministerial Conference was an early exemplar of this organizing model. Later, activists would replicate the Global Day of Action template in over a dozen cities in the early 2000s as we showed with the Prague IMF and World Bank Meetings.

As the impending US invasion of Iraq took the spotlight away from global justice in 2003, activists once again emulated the Global Day of Action template and coordinated the largest simultaneous protest event in history: the February 15 worldwide protests against the impending war in Iraq with a reported 10 to 15 million protest participants. While this global activist organizational infrastructure was solidified, the planet continued to warm and the new UN mechanisms for international climate negotiations stagnated. The groups of NGOs, environmental organizations, and scientists advocating for a sane climate policy split into radical and reformist factions as they lost patience with the institutional processes in place. In this context, the Climate Justice movement arose in conjunction with the annual UN-sponsored Conference of the Parties (COP) climate negotiation meetings.

Starting in 2005, the Climate Justice movement successfully coordinated its own Global Day of Action around reducing carbon emissions. By 2006, the Climate Justice movement reached every continent, and by 2009, the movement had clearly established itself as one of the most extensive transnational movements in the contemporary period. We analyzed eleven major climate justice annual global actions between 2005 and 2018. We demonstrated that nations with more Global Economic Justice and Anti-War movement activity were associated with greater climate justice action. These precursor movements provided the transnational activist infrastructure that would later be appropriated by the climate justice campaigns. The continuing spread of global internet technology to poorer countries in the 2000s and 2010s was also a major indicator of a country's participation in climate justice campaigns as was the ecological threat of monster storms and sea level rises for countries with large populations living just above sea level. We conclude that the Climate Justice movement's inclusiveness and extensiveness can lead it to be the fulcrum that links other progressive movements for global change in the 2020s and beyond. Indeed, the continuing ecological climate threat has the capacity of coalescing the current family of progressive movements into an enduring and capable global-level alliance (Almeida 2019a).

The World Social Forum and Coalitions of the Future

Chapter 4 showcased the concept of world revolutions (clusters of revolts and revolutions in world history that shaped the evolution of economic and political institutions in the Europe-centered world-system and that may have played a similar role in the East Asian world-system). The World Social Forum (WSF) was introduced as the most significant effort to put together a rainbow coalition for a global progressive movement in the World Revolution of 20xx. Limits of horizontal organizing were also discussed in the context of acknowledging the indispensable role states, labor unions, and political parties play in coordinating large numbers of people.

We report the results of four surveys of attendees carried out at World Social Forum venues in Porto Alegre, Brazil, in 2005; Nairobi, Kenya, and Atlanta, Georgia, in 2007; and Detroit, Michigan, in 2010. Asking attendees about active involvement in a large set of movement themes, we were able to use the results to study the relationships among movements. Formal network analysis of these network connections reveals a set of intermovement linkages that are rather stable over time and across the different venues that were studied. We found that there was a small cluster of movement themes at the center of the intermovement network, and environmentalism was among these. This implies support for our contention that environmentalism and the Climate Justice movement could be the fulcrum of a global progressive effort to transform the modern world-system into a democratic, sustainable, and collectively rational global commonwealth. We review the sociological literature on coalition formation and consider the pros and cons of environmentalism, a revitalized labor movement, feminism, and the peace movement as potential fulcra for an organized global progressive alliance that could make a difference in world politics. We also review some recent proposals for a renewed effort to construct an ethical and philosophical basis for an integrated humanistic progressive global civilization. And we discuss organizational proposals that are intended to help resolve tensions within the culture of the New Global Left between horizontalism and the need for greater global coordination. The success of the Global Left in the world revolution of 20xx will require the articulation of a wide variety of groups, including feminists, labor activists (the precariat and unions in both the Global North and the Global South), peace activists, sexual-orientation rights activists, anti-racists, human rights activists, indigenous peoples, and several other progressive movements. An umbrella intersectional alliance perspective theorized in terms of humanism, race, class, gender, sexual orientation, and environmental stewardship, and focusing on the crisis of anthropomorphic climate change, may be our best hope for solidifying such a rainbow coalition.

The Evolution of the Global Right

Chapter 5 discusses the evolution of the structure of world politics since the early nineteenth century in terms of changing interactions between the Global Right, the Global Left, and Centrist Liberalism and their adaptations to the changing institutional context of the modern world-system– the emergence and deepening of global capitalism, the rise and fall of hegemonic core states, and other similar transformations. The structure of world politics–the geoculture—is understood as part of a larger emergent global culture in which the civilizational and national cultures are converging with respect to institutionalized assumptions about what exists (ontology) and what is good (ethics, morality, legality, etc.). The tripartite structure of the geoculture is largely a product of the World Revolution of 1789, especially the French Revolution. Chapter 5 summarizes the social science literature on twentieth-century fascism and compares it with the right-wing populist and neofascist movements, parties, and regimes that have emerged in the twenty-first century. We also discuss the coevolution of logics of accumulation in the context of their world historical evolution. Fascism and neofascism are understood as hybrid combinations of the tributary mode of accumulation and capitalism, and the possibility of the emergence of a global tributary police state is contemplated but is seen as unlikely. Twenty-first century authoritarian populism and neofascism are driven by nationalism, xenophobia, racism, and economic decline in the core industrial centers of the Global North and neoliberalism in the Global South. Because of its nativist and nationalist features, there may be limits to its growth as a potent transnational movement, but a new transnationalist white identitarian movement is trying to overcome this liability (Spektorowski 2016).

Global Change and Social Movements

We have shown throughout these pages that global forces play a crucial role in shaping local, national, and transnational movements. We have attempted to overcome the bias of social movement theory that focuses largely on national-level movements and meso-level organizational factors below the global level of social life. Though chapter 5 contemplates some unhappy possibilities regarding the trajectory of global politics in the next several decades, we contend that early twenty-first-century activists have established an enduring, progressive, international organizational infrastructure that can move global society in a good direction. It consists of a loosely coupled network of activists and organizational ties often coordinated through the ICT channels that the Global Justice and Anti-War movements initiated. While we have witnessed several setbacks in the expansion of positive transnational social movement outcomes in recent years, the global activist infrastructure has been successfully appropriated by the Climate Justice movement. Chapter 5 also discusses conditions under which the crisis of the twenty-first century could result in another systemic cycle of global capitalism or a transformation to a post-capitalist democratic and sustainable global commonwealth. It is noted that these alternatives will be partly shaped by the strategic actions of progressive actors and by the timing of calamities. A perfect storm of simultaneous financial, environmental, and health crises could produce an opening for progressive systemic transformation.

The major tasks ahead rely on both increased and better individual organizing and a willingness to build a unifying organization that will bring progressive sectors into a long-lasting and capable international coalition. Transnational movement organizers have worked hard to produce tremendous mobilization events that involve large numbers of demonstrators in many locations simultaneously. This is a great tactic that has had big effects on political discourse, but the ability to engage "flash activism" in community-building and sustained campaigns to change institutions is quite limited. Strategic projects should balance digital activism with face-to-face forms of solidarity and organizational capacity building across neighborhoods, communities, and nations in order to build the components of an enduring and efficacious transnational movement and a global progressive organization. This pertains to why the World Social Forum process proved so critical. It moved beyond the one-day events of the Global Days of Action template to facilitate intense in-person discussions and confrontations at the real-time meetings of large numbers of progressive activists. In the post-COVID-19 world, distance interaction on the internet will be an even more important space for global organizing. But a reinvented Social Forum process for face-to-face interactions and the formation of a formally organized global party/network of the Left will also be needed.

A unifying framework has the challenge of balancing cohesion and inclusion. Beyond the relatively successful framing strategies of the World Social Forum and the Climate Justice movement, a broader coalitional grouping will be necessary to sustain a progressive transnational movement in the 2020s. One of the more successful models in creating alliances in the twenty-first century has employed the school of thought around intersectionality. The intersectionality framework was used to mobilize the three Women's Marches in the United States (and around the world) against Trumpism between 2017 and 2020—the largest simultaneous protests in US history. With the United States's heterogeneous population characteristics (in terms of race, ethnicity, religion, and immigration status), the intersectional template provides a unifying umbrella for both national and transnational social movements. It is a model that appeals across race, class, national origin, and gender categories and identities, while explicitly acknowledging and contesting multiple forms of oppression (Chun, Lipsitz, and Shin 2013; Collins and Bilge 2016; Luna 2016; Rojas 2017; Terriquez et al. 2018; Crenshaw 2020). The comparative world-system perspective when combined with an intersectionality approach provides a powerful framework for understanding and acting to transform world society.

The perfect storm that might provide an opening for transformational global social change mentioned in chapter 5 may have arrived with the global COVID-19 pandemic. A widely circulated video purporting to be a message from the COVID-19 virus to a humanity locked down in an effort to save lives notes how clean the air is (https://www.filmsforaction.org/watch/a-letter-from-the-virus-listen/). Global indigenistas will see this as a wakeup call from Pachamama (Mother Earth). At the same time, such a wakeup call would begin by explicitly recognizing how COVID-19 disproportionately harms the most vulnerable groups across the planet in a left-green alliance.

The progressive rainbow coalition of the next several decades will benefit from leadership by activists from globally marginalized and excluded communities and social sectors of the kind that the World Social Forum was designed to promote. These movement leaders from the Global South and from marginalized groups in the Global North will be able to spearhead a flexible intersectional alliance that acknowledges and contests the different forms of stratification and privilege that are institutionalized in the current global system. This should include a global partnership of groups across race, class, gender, religious/spirituality, and sexual orientation lines and a global justice perspective that focuses on both global inequalities and inequalities within societies of the Global North and the Global South. Scholar-activists are proposing new organizational models for a needed global party/network as discussed in chapter 5. This effort should strongly include the intersectionality perspective championed by the World Social Forum. The transnational progressive intersectional alliance should also build internal cohesion by mobilizing against the xenophobia and climate change denialism of right-wing populism and neofascism. Along with the energy of participatory democracy, a potent instrument should also include a supportive and enduring infrastructure composed of progressive states, parties, and formal organizations such as labor unions, progressive NGOs, and progressive faith-based institutions. These will provide the basic building blocks for a mobilized alternative to climate collapse and demagogic authoritarianism as we enter the second quarter of the twenty-first century.

Notes

....

Introduction

1. World revolutions are periods in global history in which local and national rebellions and revolutions break out across the world-system. In the early centuries the rebels were unaware that their rebellions were occurring in the same time periods but the home offices of colonial empires knew and had to respond, and thus these collective behavior events were indirectly linked with one another through the hierarchical structures of the world-system.

2. The terminology of the world-system perspective divides the Global South into the periphery and the semiperiphery. This turns out to be an important distinction for comprehending political developments in the Global South. Activists from the semiperiphery have been far more likely to participate in the Social Forum process, and activists from the periphery have been much more critical of international political organizations than those from either the Global North or the semiperiphery (Chase-Dunn et al. 2008).

3. Some definitions of global civil society exclude advocates of armed struggle, but these should be included because they sometimes have important effects on world order. Some analysts of transnational social movements contend that alliances among national civil societies are not sufficiently integrated enough to merit the term *global civil society* (e.g., Keck and Sikkink 1998). We see a long-term trend in which the size and level of integration among actors focusing on the global field of politics has grown, but we agree that this is still only a small part of world politics that is mainly composed of a collection of largely autonomous local and national political communities. We focus on global civil society because we think it is part of a long-term trend toward the formation of global citizenship but we also acknowledge that this process has a very long way to go, and that waves of nationalism and deglobalization such as the one that is now on the rise slow down the long-term trend.

Chapter 1. Social Movements and Collective Behavior in History and Prehistory

1. We use the term *polity* to generally denote a spatially bounded realm of sovereign authority such as a band, tribe, chiefdom, state, or empire. Small-scale polities are bands, tribes, and chiefdoms.

2. Use of the word *evolution* still requires explanation. We mean long-term patterned change in social structures, especially the development of complex divisions of labor and hierarchy. We do not mean biological evolution, which is a very different topic, and neither do we mean "progress." Defining what has happened to social structures since the Stone Age as improvement (or decadence) is not a necessary exercise for the scientific description and explanation of these changes. Recall that, in the introduction, we identify our general theoretical perspective as "institutional materialism," an approach that combines structural functionalism with conflict theory. We do not reject functionalism but see it in a context of historical struggles involving the efforts of elites who are in competition with one another in a context of rebellions and resistance from below. This is world-system Marxism.

3. We do not agree with those critics of Eurocentrism who argue that the values of the European Enlightenment and the inventions of European social science are all wrong because they were produced in the context of the rise and consolidation of European global hegemony. Some of the ideas are bad but some are very good.

4. When social movements are defined as excluded social groups using nonconventional strategies to seek social change by targeting elites, institutions, and governments, the term "excluded social groups" feeds the idea of collective action from below that we are arguing represents a conceptual limitation for prehension of many premodern social movements. See Mora et al. (2017) and Almeida (2019) for the excluded groups definition of movements in the contemporary era.

5. This corresponds with what Tilly and Tarrow (2015) mean by "contentious politics."

6. The recent example of key factions of the Tea Party in the United States receiving sponsorship from the Republican Party and mass media is one elite-led movement (Almeida and Van Dyke 2014). Another is found in clientelist networks in Argentina led by party bosses organizing collective action on the streets (Auyero 2007).

7. Small-scale polities are bands or tribelets in which autonomous authority does not extend very far in space and does not include very large populations. Systems based on these kinds of polities are usually peopled by nomadic or sedentary hunter-gatherers or by horticulturalists who live in relatively small settlements (camps, hamlets, or villages).

8. Hunter-gatherers (foragers) are people who get most of what they eat by harvesting what nature produces naturally without much intervention intended to

increase natural productivity. They do not plant or plant only a little. They do not herd. The transition from nomadism to sedentism occurred on many regions before the emergence of horticulture. Archeologists call sedentary hunter-gatherers who live in permanent winter villages with the term Mesolithic. The transition from nomadism to sedentism was a big evolutionary change in which population density went up significantly. Precontact California was a region in which sedentary foraging polities continued to exist until the arrival of Europeans in the last few centuries.

9. A. F. C. Wallace (1965: viii) says that both the 1870 and the 1890 versions of the Ghost Dance doctrine held that "the dead were soon to return and that the white people and their culture were at the same time to be destroyed by a natural cataclysm."

10. James Fenelon (1998) contended that the Ghost Dance doctrine did not predict the disappearance or death of the Europeans, but that this was attributed to the movement by whites who were nervous about the intentions of the restless natives and intent on "culturicide." This is plausible and the different reports that are relevant are rather vague about exactly what was said or predicted to be the fate of the whites in Ghost Dance doctrine. Mooney (1965: 19) notes that there were differences among groups about this element of the doctrine and that Wovoka and many other adherents preached peaceful relations with the whites. But most of the ethnologists who studied the Ghost Dance at the time or later were rather sympathetic to the plight of the indigenes. They also were told that the doctrine included bad ends for nonbelievers and for metis (half-breeds). Mooney (1965: 227) includes an Arapahoe song about the "yellow hides" that tends to support the idea that the future utopia without sickness or death did not include a place for the Europeans. See also Ruby and Brown (1989).

11. Indeed, one of DuBois's informants said, "A white man looks at paper and talks to it and laughs." ([1939] 2007: 50).

12. Traditional shamans were called "sucking doctors" by Cora DuBois because they cured patients by removing foreign objects (bad spirits) from their bodies.

13. Chief Alexander (Sunusa) of the Upper Sacramento Wintu in Northern California convinced his ally Bogus Tom to travel to Oregon to spread the Ghost Dance word (DuBois [1939] 2007).

14. The Amerindian notion of sequential worlds and transformations is quite like Millenarianism.

15. Cargo cults were Melanesian Millenarian movements encompassing a diverse range of practices and that occurred in the wake of contact with the commercial networks of colonizing European polities. The name derives from the belief that various ritualistic acts will lead to a bestowing of material wealth ("cargo"). Worsley (1968) saw the Millenarianism of the cargo cults as having

been borrowed from Christian missionaries, whereas Lawrence (1964) contended that Millenarianism was part of the precontact Melanesian culture.

16. This reminds us of Karl Marx's (1844) poignant remark that religion is the sigh of the oppressed creature, the heart of a heartless world, and the soul of soulless conditions.

17. Spier (1935: 10) also notes that the Modoc of Northern California were known to have engaged in a version of the Prophet Dance well before the emergence of the Ghost Dance in Western Nevada in 1870. He also contends that the 1870 Ghost Dance had little or nothing to do with the vicious war that broke out in 1872 between the Modoc and the US Army, though that seems difficult to believe.

18. Spier (1935: 12) contends that the proselytizing aspect of the Prophet Dance was probably not due to exposure to Christianity.

19. Scale is relative. Chaco Canyon had about ten thousand residents. The largest settlements in precontact California had about two thousand. They are both small compared to cities of the twenty-first century but were very different scales compared with one another.

20. Regarding scale, see note 19 above.

21. Ritual blood sacrifices and human sacrifices were found in many Amerindian societies as well as in other complex chiefdoms and early states. Ritual human sacrifice is a dramatic demonstration of power that was more functional before states and empires became fully institutionalized.

22. James Scott (2017) emphasizes the vulnerability of reliance on monocropping and high population density to drought, famine, epidemic diseases, and the frequent collapses of these early cities and states, but each increase in the scale and complexity of human polities has been followed by downswings, but, at least so far, these have always been followed by recoveries and eventual upswings.

23. Recent archaeological findings from Turkey, five hundred miles north of Mesopotamia, have uncovered evidence of child sacrifice about five hundred years earlier than the Ur-3 evidence (Hignett 2019).

24. See Chase-Dunn and Lerro (2014) for further discussions of territorial upsweeps and marcher states.

25. The Axial Age was a period in the middle of the first millennium BCE in which innovative thinkers and philosophers emerged in several world regions.

26. While reading about human sacrifice in Mayan religions, co-author Chase-Dunn arrived at a village in the mountains above Guatemala City to witness an Easter Parade in which the modern Mayans were dressed up as Roman soldiers to escort Jesus on his way to the cross.

27. Little wonder that the Chinese Communist Party is made very nervous by Falun Gong, a religious movement that is similar in many respects to the Taiping.

Chapter 2. Resistance to Neoliberalism in the Global North and South

1. The extension of rights has also played a major role in mass mobilization, especially in nondemocracies (Schock 2005; Almeida 2008). Nonetheless, opposition to neoliberalism is largely driven by threats to economic livelihoods and social citizenship rights.

2. These data do not include the massive outbreaks of austerity and economic protests in late 2019 occurring in Haiti, Iraq, Iran, Colombia, Pakistan, Lebanon, Ecuador and Chile.

3. One obvious exception to these claims would be indigenous and rural peoples that were displaced from their ancestral lands for mega-development and infrastructural projects (Scott 1998).

4. There is obviously a great level of variation in the bureaucratization of national governments and economies in the developing world, and this variation does condition the degree of collective action. Even with the high level of bureaucratic variation between countries in the Global South, all or almost all nations have more formal procedures and organizations run along bureaucratic lines than before the period of state-led development.

5. It is also important to recognize ongoing everyday forms of resistance to neoliberalism that escapes coverage from media outlets, especially in terms of rural groups battling extractive industries and other struggles.

6. Data on the Combo Fiscal protests in figure 2.5 and table 2.1 come from Alvarado Alcázar and Martínez Sánchez (2018).

7. The demonstrators in the massive anti-neoliberal protests in Ecuador and Lebanon in October of 2019 used the highway roadblock as a principal tactic.

Chapter 3. Transnational Movements

1. August 2019 tied for the second hottest month ever recorded, September 2019 tied with September 2015 for the hottest month ever recorded.

2. African countries are underrepresented in this early phase of the Global Justice movement but are increasingly participating in the Climate Justice movement.

3. See World Bank Data at https://data.worldbank.org/indicator/it.NET .user.ZS.

4. The eleven climate justice campaigns are those listed in table 3.1 and occurred in 2005, 2006, 2007, 2008, 2009, 2010, 2011, 2012, 2014, 2015, and 2018.

5. See Timothy Gardner, "Climate Activists Hope to Bring U.S. Capital to Standstill on September 23." Reuters, September 11, 2019. https://www.reuters

.com/article/us-usa-climatechange-washingtondc/climate-activists-hope-to-bring
-u-s-capital-to-standstill-on-september-23-idUSKCN1VW2B4. A recent coalition
between the Central Única dos Trabalhadores (CUT) and the Fridays for Future
movement in Brazil in late 2019 offers another example of cross-sectoral alliances
as the CUT has called a general strike along with the youth climate strike on September 20, 2019, see "Anuncian marchas contra Bolsonaro el 20 de septiembre
en Brasil." Telesur, September 10, 2019. https://www.telesurtv.net/news/brasil
-central-trabajadores-anuncio-marchas-contra-bolsonaro-20190910-0017.html.

Chapter 4. The New Global Left and the World Revolution of 20xx

1. Samir Amin died on August 12, 2018. His call for the establishment of
a global fifth international that would coordinate and support progressive social
movements was published in 2019 (Amin 2019) as was a forum addressing his
proposal (Gills and Chase-Dunn 2019).

2. Anti-systemic movements include a diverse "family of movements" working to advance greater democracy and equality. According to Wallerstein, "to be
anti-systemic is to argue that neither liberty nor equality is possible under the existing system and that both are possible only in a transformed world" (1990: 36).

3. This demise in the 2010s includes major electoral defeats of left parties
in Argentina and El Salvador but also extra-constitutional interventions against
leftist governments in Brazil, Honduras, and Paraguay, as well as shifts toward
authoritarianism in Nicaragua and Venezuela.

4. Our categorization of reformist and anti-systemic regimes in Latin America from 1959 to 2012 is contained in the appendix to Chase-Dunn, Morosin, and
Alvarez (2015) which is available at http://www.irows.ucr.edu/cd/appendices
/pinktide/pinktideapp.htm.

5. The Charter of the World Social Forum discourages participation by those
who attend as representatives of organizations that are engaged in, or that advocate, armed struggle; nor are governments, confessional institutions, or political
parties supposed to send representatives to the WSF. See the World Social Forum
Charter of Principles.

6. These bridges were studied in a local context by Carroll and Ratner
(1996).

7. While human rights are a very central movement theme in the network
of Global Justice movements, the Indigenist Rights movement contests the version
of human rights that is enshrined in the United Nations Universal Declaration of
Human Rights of 1948. The indigenistas stress the importance of community
rights over the rights of individuals and the idea that "Mother Earth" has rights.
These contentions have been shared by the many activists who sympathize with,

and identify with, indigenous peoples (Chase-Dunn, Fenelon, Hall, Breckenridge-Jackson, and Herrera 2020).

8. Some well-known examples are the Rosa Luxemburg Foundation, the Third World Forum, the Centre for Civil Society, Development Alternatives with Women for a New Era, and Focus on the Global South.

9. Our project web page contains the survey instruments that were used at four Social Forum venues from 2005 to 2010. http://www.irows.ucr.edu/research/tsmstudy.htm.

10. What we call "movement themes" include both ideological constellations (e.g., anarchism, communism, etc.) and topical issues. The latter groupings of social movement organizations around their goals have been called "social movement industries" (Zald and McCarthy 1987; Snow and Soule 2010: 152).

11. The UCINet QAP routine produces a Pearson's r correlation coefficient that shows the degree of similarity between two dichotomized affiliation network matrices. The Pearson's r coefficient varies from -1 (a perfectly negative linear relationship between two variables) and +1 (a perfectly positive linear relationship). The Pearson's r correlation coefficient between the USSF 2007 and the USSF 2010 movement affiliation matrices was 0.74. This is a rather strong positive correlation and is slightly larger than what was found between the World Social Forum meeting in Nairobi and the US Social Forum meeting in Atlanta, which was 0.71 (Chase-Dunn and Kaneshiro 2009). It is unsurprising that the Atlanta USSF network would be more similar to the Detroit USSF than it would be to the Nairobi WSF, but the surprise is that the national and global affiliation matrices are so similar.

12. The BRICS are Brazil, Russia, India, China, and South Africa.

13. Despite these anti-statist stances, the World Social Forum received important support from Pink Tide regimes in Latin America.

14. All of the three hegemons of the modern world-system (the Dutch, the British, and the United States) were former semiperipheral states and the Chinese and Russian revolutions of the twentieth century occurred in semiperipheral countries.

15. Zald and McCarthy (1987: 168–169) discuss how competition between radical and reformist movement organizations is exacerbated by the greater likelihood that the reformists will be granted legitimacy by authorities, but they also mention "the functions of the radical fringe."

16. Marx-type unrest occurred among "the working classes that have been formed in those East and South Asian countries that are undergoing economic transformations. These new working classes are putting forth offensive demands and, in doing so, have made East and South Asia global centers of labor unrest (Karatasli et al. 2014).

17. Polanyi-type "protests belong to working classes that are currently being unmade in one way or another: Public-sector workers are losing their previously gained rights and privileges due to austerity politics. Workers are resisting the closing down of factories, mines, or state-owned enterprises, and are protesting the restructuring of the pay scales that jeopardize overtime pay, bonuses and special allowances" (Karatasli et al. 2014; see also Burawoy 2012; and figures 2.1 and 2.2 in chapter 2).

18. "We find protests of working classes being unmade in declining centers of production, and protests of working classes being made in rising centers of production, with localized mixes of the two found across the world economy" (Karatasli et al. 2014).

19. "These are workers who, because they are superfluous to the needs of existing capital, have extremely irregular employment and thus demand primarily 'more jobs'" (Karatasli et al. 2014).

20. Mike Davis (2006) has also pointed to the significance of the huge portion of humanity that has been bypassed by the capitalist accumulation process.

21. Other research lends support to the notion that the Occupy movement was strongly supported by "graduates without a future" (Milkman, Luce, and Lewis 2013; Curran, Schwarz, and Chase-Dunn 2014).

22. Guy Standing (2011, 2014), the economist who has written about the precariat and is, with Mason, a consultant to Jeremy Corbin, has a rather different view of tweeting. He thinks it makes people stupid because it reduces their attention spans.

23. Prefigurationism is the idea that small groups and communities can intentionally organize social relations in ways that can provide the seeds of transformation to a more desirable form of future human society.

Chapter 5. Evolution of the Global Right in the Geoculture

1. The term hegemony is used here in the world-system sense of a predominant concentration of global economic power is a single core state. Global governance in the modern world-system has mainly been the result of the rise and fall of a succession of hegemonic states (the Dutch in the seventeenth century, The British in the nineteenth century and the United States in the twentieth century) (Wallerstein 1984). Ideological hegemony as theorized by Antonio Gramsci refers to the ideological class struggle and the power of the ruling class to impose its world-view on society. The Gramscian perspective has been extended to the global level by Robert Cox, Stephen Gill, William Carroll, and other international relations scholars. We use the term "counter-hegemonic" in the Gramscian sense above.

2. Kondratieff Waves are long (forty to sixty year) business cycles that have operated in the modern world-system since at least 1790 (Goldstein 1988). A-phases are periods of economic expansion in which capitalists invest in labor-saving technologies because workers and their organizations successfully press for higher wages. B-phases are periods of slower economic growth. Paul Mason (2015) explains how the policies of neoliberal globalization and the weakness of the labor movement interrupted the arrival of another Kondratieff A-phase that would normally have occurred near the end of the twentieth century.

3. World-system theorists focus on global inequalities, but their terminologies have been somewhat different. Samir Amin and Andre Gunder Frank talked about "center" and "periphery." Immanuel Wallerstein proposed a 3-tiered structure with an intermediate semiperiphery between the core and the periphery, and he used the term "core" to suggest a multicentric region containing a group of states rather than "center," which implies a hierarchy with a single peak. When the world-system perspective emerged the focus on the noncore (periphery and semi-periphery) was called Third Worldism. Current terminology refers to the Global North (the core) and the Global South (periphery and semiperiphery). The term "noncore" refers to the semiperiphery and the periphery.

4. The world-system perspective understands the modern capitalist mode of accumulation as necessarily combining commodity production using wage labor in the core with various forms of commodity production using coerced labor (slavery, serfdom, etc) in the noncore (Chase-Dunn 1998).

5. See discussion of nationalism below.

6. The charge that radical environmentalist and advocates of deep green philosophy are misanthropists has mainly been leveled by opponents of environmentalism. Radical ecocentrists have not worked well with the environmental justice movement that focusses on how pollution, etc., mainly impacts disadvantaged populations, but they have not explicitly expressed anti-human sentiments (Woodhouse 2018).

7. In the introduction and chapter 4, we have defined and discussed the concept of world revolutions.

8. The Concert of Europe was the beginning of the erection of supranational governmental organizations, a trajectory that led to the League of Nations and then the United Nations. This two-century old process will likely continue and eventually lead to the emergence of a true global state (Chase-Dunn and Inoue 2012). The political and social nature of that eventual global state is a subject of our discussion of the future on the tributary mode of accumulation.

9. As Marx (1852) said, comparing Napoleon III with the original Bonaparte, history repeats itself, the first time as tragedy, the second time as farce.

10. In chapter 4 of *Global Formation*, Chase-Dunn (1998) contended that normative regulation based on cultural consensus was the main glue holding small-scale human societies together, but that markets and states had emerged as major forms of regulation in complex social systems, relegating normative regulation to a supportive role. See also Chase-Dunn (2014).

11. The Chinese trade-tribute structure that was predominant in the East Asian world-system was a somewhat milder form of the tributary mode in which ritual exchange played an important role in reinforcing the notion of China as the center of the human universe (Arrighi et al. 2003).

12. There were fascist movements in the United States and Britain, but they never gained much headway compared with the stronger movements in Germany, Italy, and Spain. Fascist ideology distinguished between bourgeois and proletarian states, suggesting an awareness of inequality within the core.

13. Goldfrank (1978: 78) says, "In contrast to the varieties of communist parties and states, the differences among the fascisms are mandated, as it were, by nationalist principles rather than mere adaptations to local traditions or political exigencies."

14. Even when fascism was secular it was usually formulated as a mystical essence based on either race or culture and on topophilia (sacred soil and place).

15. Nearly all emergent regimes combine some popular movement from below with elite mobilization, so this distinction is one of degree, but it is difficult to be exact about the balances of forces involved in the emergence of regimes.

16. A similar phenomenon emerged after the Mexican Revolution (1926–1929). The Cristeros were peasants in west-central Mexico who fought against the Calles government to support the church and traditional authorities and were in turn supported by the Los Angeles diocese of the Catholic Church (Davis 1992: chap. 6).

17. The emergence of transnational racial solidarities among white Europeans (Spektorowski 2016) may reduce or reconfigure this difference. See discussion below.

18. The US alt-right leader Steve Bannon's effort to lend support to the French National Front in the European parliament election of 2019 was rebuffed by Marie LePen, presumably because Bannon was seen as interfering in French politics.

19. The 2018 meeting of the World Social Forum in Mexico City focused on the plight and rights of migrants.

20. Our categorization of reformist and anti-systemic regimes in Latin America from 1959 to 2012 is contained in the Appendix to Chase-Dunn, Morosin, and Alvarez (2014) which is available at http://www.irows.ucr.edu/cd/appendices/pinktide/pinktideapp.htm.

21. Farid Alatas (2019) discusses the history of progressive Islam, which is a potential ally of the Global Left.

22. While counter-hegemonic ideologies became increasingly secular, religion continued to play an important part in politics. Mike Davis (2018) notes that Catholic political parties played an important role in undermining socialist parties in Europe in part by championing family values. And Linda Gordon's (2017) study of the movement of the Ku Klux Klan to northern US cities in the 1920s notes that the KKK championed "true Americanism," racial purity, religious intolerance and opposition to immigration.

23. The organizational structure of DiEM25 is intentionally diagonal (combining local self-organization with democratically controlled coordination. This could serve as a template for a global party network of the left (Alvarez and Chase-Dunn 2019; Patomäki 2019; see also Moghadam 2020b).

24. Valentine Moghadam (2017) shows how gender relations and women's mobilizations prior to the protest outbreaks, along with differences in political institutions, civil society, and international influences, explain most of the variance in the different outcomes of the Arab Spring.

25. BRICS is an acronym that refers to the countries of Brazil, Russia, India, China and South Africa—newly industrializing countries in the semiperiphery.

26. When President Trump declared that the United States will never be socialist, a new discussion about the meanings of socialism and democracy began in the mass media in the United States.

27. Samir Amin (2019) proposed the organization of a fifth international to support progressive national political projects in the Global South. Heikki Patomäki (2019) has a new proposal for a global political party of the Left, and Patrick Bond (2019) reviews and critiques efforts to organize at the global level that have occurred in recent decades.

28. Violent video games, movies, television, and social media abound but these are not the province of right-wing movements. Gun culture in the United States is broadly entrenched. The mass murders that have been carried out by shooters inspired by right-wing propaganda so far appear to be the work of deranged individuals rather than organized groups. Ironically this seems similar to the anarchist notion of the propaganda of the deed. Our point is that, at least so far, both domestic violence and militarism seem to be less intense than was the case with twentieth-century fascism.

29. Alberto Spektorowski (2016) claims that racial ethno-regionalism is supplanting nationalism. He argues that post-national European fascism may be the next stage in the evolution of fascism in transnational regions with a focus on preserving an "ethnic federation of European people" in the form of a "strong, dominant, and productive conglomeration" (Spektorowski 2016: 126). If he is

right, a new form of fascist transnationalism may reduce the difference in this regard between the Global Right and the Global Left discussed above.

30. An exception was Donald Trump's mention of "the Second Amendment people" during the US presidential campaign of 2016.

31. But, at least so far, lethal violence appears to have been restricted to mentally challenged individuals inspired by the rhetoric on internet image boards such as 8chan.

32. The demonstrations against "Sharia law" in San Bernardino, California, in 2018 falsely claimed that the Koran condones female genital mutilation, thereby attempting to mobilize women and feminists to support anti-Muslim and anti-immigrant causes.

References

Abouharb, M. Rodwan, and David Cingranelli. 2007. *Human Rights and Structural Adjustment*. Cambridge: Cambridge University Press.

Accornero G., and P. Ramos Pinto. 2015. "'Mild mannered'? Protest and mobilisation in Portugal under austerity, 2010–2013." *Western European Politics* 38 (3): 491–515.

Alatas, Sayed Farid. 2019. "Against the Grain: The Meaning of Progressive Islam." *The Edge Malaysia*, September 5. https://www.theedgemarkets.com /article/against-grain-meaning-progressive-islam?fbclid=IwAR0lpWLHO D7c65KU3CsEfutx0JsEWnfvf1vV2enuqPTc4aO52HiaBTcxw7g.

Aldecoa, John, C. Chase-Dunn, Ian Breckenridge-Jackson, and Joel Herrera. 2019. "Anarchism in the Web of Transnational Social Movements." *Journal of World-Systems Research* 25 (2): 373–394.

Almeida, Paul D. 2003. "Opportunity Organizations and Threat Induced Contention: Protest Waves in Authoritarian Settings." *American Journal of Sociology* 109 (2): 345–400.

Almeida, Paul D. 2007. "Defensive Mobilization: Popular Movements against Economic Adjustment Policies in Latin America." *Latin American Perspectives* 34 (3): 123–139.

Almeida, Paul D. 2008a. "The Sequencing of Success: Organizing Templates and Neoliberal Policy Outcomes." *Mobilization* 13 (2): 1655–1687.

Almeida, Paul D. 2008b. *Waves of Protest: Popular Struggle in El Salvador, 1925–2005*. Minneapolis: University of Minnesota Press.

Almeida, Paul D. 2010a. "Social Movement Partyism: Collective Action and Political Parties." In *Strategic Alliances: New Studies of Social Movement Coalitions*, ed. N. Van Dyke and H. McCammon, 170–196. Minneapolis: University of Minnesota Press.

Almeida, Paul D. 2010b. "Globalization and Collective Action." In *Handbook of Politics*, ed. Kevin Leicht and J. Craig Jenkins, 205–226. New York: Springer.

Almeida, Paul D. 2012. "Subnational Opposition to Globalization." *Social Forces* 90 (4): 1051–1072.

Almeida, Paul D. 2014. *Mobilizing Democracy: Globalization and Citizen Protest*. Baltimore: Johns Hopkins University Press.

Almeida, Paul D. 2016. "Social Movements and Economic Development." In *The Handbook of Development*, ed. G. Hooks, 228–250. Berkeley: University of California Press.

Almeida, Paul D. 2018. "The Role of Threat in Collective Action." In *Wiley-Blackwell Companion to Social Movements*, 2nd ed., ed. D. Snow, S. Soule, H. Kriesi, and H. McCammon, 43–62. Oxford: Wiley-Blackwell.

Almeida, Paul D. 2019a. "Climate Justice and Sustained Transnational Mobilization." *Globalizations* 16 (7): 973–979.

Almeida, Paul D. 2019b. *Social Movements: The Structure of Collective Mobilization*. Berkeley: University of California Press.

Almeida, Paul D. 2020. "USA-Country Report." In *Protest for a Future II: Composition, Mobilization, and Motives of the Participants in Fridays For Future Climate Protests on 20–27 September, 2019, in 19 Cities around the World*, ed. Joost de Moor, Katrin Uba, Mattias Wahlström, Magnus Wennerhag, and Michiel De Vydt, 247–253. doi:10.17605/OSF.IO/ASRUW

Almeida, Paul D., and Christopher Chase-Dunn. 2018. "Globalization and Social Movements." *Annual Review of Sociology* 44: 189–211.

Almeida, Paul D., and Mark I. Lichbach. 2003. "To the Internet, from the Internet: Comparative Media Coverage of Transnational Protest." *Mobilization* 8 (3): 249–272.

Almeida, Paul D., and Nella Van Dyke. 2014. "Social Movement Partyism and the Rapid Mobilization of the Tea Party." In *Understanding the Tea Party Movement*, ed. D. Meyer and N. Van Dyke, 55–72. London: Ashgate.

Alvarado Alcázar, Alejandro, and Gloriana Martínez Sánchez. 2018. "La huelga general contra La Reforma Fiscal en Costa Rica." Instituto Investigaciones Sociales, Universidad de Costa Rica.

Alvarenga Venutolo, Patricia. 2005. *De Vecinos a Ciudadanos: Movimientos Comunales y Luchas Cívicas en la Historia Contemporánea de Costa Rica*. San José: Editorial de la Universidad de Costa Rica.

Alvarez, Rebecca. 2019. *Vigilante Gender Violence and the Sociocultural Evolution of Gender Inequality*. London: Routledge.

Alvarez, Rebecca, and Christopher Chase-Dunn. 2019. "Forging a Diagonal Instrument for the Global Left: The Vessel." *Journal of World-Systems Research* 25 (2). doi:10.5195/jwsr.2019.947.

Amenta, Edwin, Kenneth Andrews, and Neal Caren. 2018. "The Political Institutions, Processes, and Outcomes Movements Seek to Influence." In *The Wiley-Blackwell Companion to Social Movements*, ed. D. Snow, S. Soule, H. Kriesi, and H. McCammon, 449–465. Oxford: Blackwell.

Amin, Samir. 1976. *Unequal Development: An Essay on the Social Formations of Peripheral Capitalism*. New York: Monthly Review Press.

Amin, Samir. 1980a. *Class and Nation, Historically and in the Current Crisis*. New York: Monthly Review Press.

Amin, Samir. 1980b. "The Class Structure of the Contemporary Imperialist System." *Monthly Review* 31 (8): 9–26.

Amin, Samir. 1990a. *Delinking: Towards a Polycentric World*. London: Zed Books.

Amin, Samir. 1990b. "The Social Movements in the Periphery: An End to National Liberation?" In *Transforming the Revolution: Social Movements and the World-System*, ed. S. Amin, G. Arrighi, A. Gunder Frank, and I. Wallerstein, 96–138. New York: Monthly Review Press.

Amin, Samir. 2008. "Towards the Fifth International?" In *Global Political Parties*, ed. Katarina Sehm-Patomaki and Marko Ulvila, 123–143. London: Zed Books.

Amin, Samir. 2019. "Letter of Intent for an Inaugural Meeting of the International of Workers and Peoples." *Journal of World-Systems Research* 2 (19). http://jwsr.pitt.edu/ojs/index.php/jwsr/article/view/960.

Anderson, David G. 1994. *The Savannah River Chiefdoms: Political Change in the Late Prehistoric Southeast*. Tuscaloosa, AL: University of Alabama Press.

Anderson, E. N. 2019. *The East Asian World-System: Climate and Dynastic Change in the East Asian World-System*. Berlin: Springer Verlag.

Anderson, Perry. 1974. *Lineages of the Absolutist State*. London: New Left Books.

Anderson, Perry. 2005. *Spectrum*. London: Verso.

Andrews, Kenneth. 2004. *Freedom Is a Constant Struggle: The Mississippi Civil Rights Movement and Its Legacy*. Chicago: University of Chicago Press.

Anheier, Helmut and Hagai Katz. 2003. "Mapping Global Civil Society." In *Global Civil Society*, ed. Mary Kaldor, Helmut Anheier, and Marlies Glasius, 241–258. New York: Oxford University Press.

Anner, Mark. 2011. *Solidarity Transformed: Labor Responses to Globalization and Crisis in Latin America*. Ithaca, NY: Cornell University Press.

Arce, Moises, and Wonik Kim. 2011. "Globalization and extra-parliamentary politics in an era of democracy." *European Political Science Review* 3 (2): 253–278.

Arce, Moises, and Jorge Mangonnet. 2013. "Competitiveness, partisanship, and subnational protest in Argentina." *Comparative. Political Studies*. 46 (8): 895–919.

Arce, Moises, and Roberta Rice. 2009. "Societal protest in post-stabilization Bolivia." *Latin. American Research Review* 44 (1): 88–101.

Armbruster-Sandoval, Ralph. 2005. *Globalization and Cross-Border Labor Solidarity in the Americas: The Anti-Sweatshop Movement and the Struggle for Social Justice*. New York: Routledge.

Arnold, Jeanne E., ed. 2004. *Foundations of Chumash Complexity. Perspectives in California Archaeology*, vol. 7. Los Angeles: Cotsen Institute of Archaeology, University of California-Los Angeles.

Arrighi, Giovanni. 1994. *The Long Twentieth Century*. London: Verso.

Arrighi, Giovanni. 2009. *Adam Smith in Beijing*. London: Verso.

Arrighi, Giovanni, Takeshi Hamashita, and Mark Selden. 2003. *The Resurgence of East Asia: 500-, 150- and 50-Year Perspectives*. London: Routledge.

Arrighi, Giovanni, T. K. Hopkins, and I. Wallerstein. 2012. *Anti-Systemic Movements*, 2nd ed. London: Verso.

Auyero, Javier. 2001. "Glocal Riots." *International Sociology* 16 (1): 33–53.

Auyero, Javier. 2002. *La Protesta: Retratos de la Beligerancia Popular en la Argentina Democrática*. Buenos Aires: Libros del Rojas (Universidad de Buenos Aires).

Auyero, Javier. 2007. *Routine Politics and Violence in Argentina: The Gray Zone of State Power*. Cambridge: Cambridge University Press.

Auyero, Javier, and Debora Alejandra Swistun. 2009. *Flammable: Environmental Suffering in an Argentine Shantytown*. Oxford: Oxford University Press.

Babb Sarah. 2005. "The social consequences of structural adjustment: recent evidence and current debates." *Annual Review of Sociology* 31: 199–222.

Babb, Sarah. 2013. "The Washington Consensus as Transnational Policy Paradigm: Its Origins, Trajectory, and Likely Successor." *Review of International Political Economy* 20 (2): 268–297.

Bahro, Rudolph. 1980. *The Alternative in Eastern Europe*. London: New Left Books.

Baker, Andy. 2009 *The Market and the Masses in Latin America: Policy Reform and Consumption in Liberalizing Economies*. Cambridge: Cambridge University Press.

Baker, Andy. 2014. *Shaping the Developing World: The West, the South, and the Natural World*. Los Angeles: Sage.

Baker, David. 2014. *Schooled Society: The Educational Transformation of Global Culture*. Palo Alto: Stanford University Press.

Bandy, Joe, and Jackie Smith. 2005. *Coalitions Across Borders: Transnational Protest and the Neoliberal Border*. Lanham, MD: Rowman & Littlefield.

Bano, Masooda. 2012. *Breakdown in Pakistan: How Aid Is Eroding Institutions for Collective Action*. Palo Alto: Stanford University Press.

Barchiesi, Franco. 2011. *Precarious Liberation: Workers, the State, and Contested Social Citizenship in Postapartheid South Africa*. Albany: State University of New York Press.

BBC (British Broadcasting Corporation). 2012. "Sudanese Newspaper Editor Dismisses Anti-Austerity Protests." *BBC Monitoring Middle East*, July 15.

Bean, Anderson. 2017. "Popular Power, Agency, and Communes in Venezuela." PhD Diss., George Mason University.

Bean, Lowell John. 1974. *Mukat's People. The Cahuilla Indians of Southern California*. Berkeley: University of California Press.

Beck, Colin J. 2011. "The World Cultural Origins of Revolutionary Waves: Five Centuries of European Contestation." *Social Science History* 35 (2): 167–207.

Beck, Ulrich. 2007. "The Cosmopolitan Condition: Why Methodological Nationalism Fails." *Theory, Culture & Society* 24 (7–8): 286–290.

Beissinger, Mark, and Gwendolyn Sasse. 2014. "An End to 'Patience'? The Great Recession and Economic Protest in Eastern Europe." In *Mass Politics in Tough Times*, ed. N. Bermeo and L. Bartels, 334–370. Oxford: Oxford University Press.

Béjar, Sergio, and Juan Andrés Moraes. 2016. "The International Monetary Fund, party system institutionalization, and protest in Latin America." *Latin American Politics and Society* 58 (2): 26–48.

Bello, Walden. 2002. *Deglobalization*. London: Zed Books.

Benjamin, Medea, and Andrea Freedman. 1989. *Bridging the Global Gap: A Handbook to Linking Citizens of the First and Third Worlds*. Cabin John, MD: Seven Locks Press.

Bennhold, Katrin. 2018. "Germany's Far Right Rebrands: Friendlier Face, Same Doctrine." *New York Times*, January 27, p. A1. https://www.nytimes.com /2018/12/27/world/europe/germany-far-right-generation-identity.html.

Berazin, Mabel. 2009. *Illiberal Politics in Neoliberal Times: Culture, Security and Populism in the New Europe*. Cambridge: Cambridge University Press.

Berazneva, Julia, and David R. Lee. 2013. "Explaining the African food riots of 2007–2008: An empirical analysis." *Food Policy* 39: 28–39.

Berberoglu, Berch, ed. 2020. *Crisis of Neoliberalism and the Global Rise of Authoritarianism in the 21st Century*. London: Routledge.

Berezin, Mabel. 2009. *Illiberal Politics in Neoliberal Times: Cultures, Security, and Populism in a New Europe*. Cambridge: Cambridge University Press.

Bhavnani, Kum-Kum, John Foran, Priya A. Kurian, and Debashish Munshi. 2019. *Climate Futures: Reimagining Global Climate Justice*. London: Zed Books.

Blaazer, David. 1992. *The Popular Front and the Progressive Tradition*. Cambridge: Cambridge University Press.

Blasi, Anthony D. 1988. *Early Christianity as a Social Movement*. New York: Peter Lang

Blumer, Herbert. 1951. "Collective Behavior." *New Outline of the Principles of Sociology*, ed. A. M. Lee, 166–222. New York: Barnes and Noble.

Bob, Clifford. 2005. *The Marketing of Rebellion: Insurgents, Media, and International Activism.* New York: Cambridge University Press.

Bob, Clifford. 2012. *The Global Right Wing and the Clash of World Politics.* Cambridge: Cambridge University Press.

Bobo, Lawrence. 2017. "Racism in Trump's America: reflections on culture, sociology, and the 2016 US presidential election." *British Journal of Sociology* 68 (Suppl. 1): S86–104.

Boehm, Christopher. 1999. *Hierarchy in the Forest.* Cambridge, MA: Harvard University Press.

Bond, Patrick. 2006. *Looting Africa: The Economics of Exploitation.* London: Zed Books.

Bond, Patrick. 2012. *Politics of Climate Justice: Paralysis Above, Movement Below.* Cape Town: University of Kwa Zulu Natal Press.

Bond, Patrick, ed. 2013. *BRICs in Africa: Anti-Imperialist, Sub-Imperialist of in Between?* Centre for Civil Society: University of Kwa-Zulu-Natal.

Bond, Patrick. 2018. "African Uprisings, Labour and Ideology in an Era of Renewed Economic Crisis: The Case of South Africa." In *Government-NGO Relationships in Africa, Asia, Europe and MENA*, ed. R. Marchetti, 25–47. London: Routledge.

Bond, Patrick. 2019a. "Truncated 21st-Century Trajectories of a Fifth International." *Globalizations* 16 (7): 1043–1052.

Bond, Patrick. 2019b. "Neoliberalism, State Repression and the Rise of Social Protest in Africa." In *The Palgrave Handbook of Social Movements, Revolution, and Social Transformation*, ed. B. Berberoglu, 213–231. New York: Palgrave Macmillan.

Boswell, Terry, and Christopher Chase-Dunn. 2000. *The Spiral of Capitalism and Socialism: Toward Global Democracy.* Boulder, CO: Lynne Rienner.

Bowles, Samuel, and Herbert Gintis 2013 A Cooperative Species: Human Reciprocity and Its Evolution, Princeton, NJ: Princeton University Press.

Branch Adam, and Zacharia Mampilly. 2015. *Africa Uprising: Popular Protest and Political Change.* London: Zed Books.

Bratton, Michael, and Nicholas van de Walle. 1997. *Democratic Experiments in Africa: Regime Transitions in Comparative Perspective.* Cambridge: Cambridge University Press.

Braudel, Fernand. 1984. *Civilization and Capitalism, 15th–18th Century Vol. 3: The Perspectives of the World.* New York: Harper and Row.

Brecher, Jeremy. 2015. *Climate Insurgency: A Strategy for Survival.* Boulder, CO: Paradigm.

Brenner, Robert. 2002. *The Boom and the Bubble: The U.S. in the World Economy.* London: Verso.

Brookes, Marissa. 2013. "Varieties of Power in Transnational Labor Alliances: An Analysis of Workers' Structural, Institutional and Coalitional Power in the Global Economy." *Labor Studies Journal* 38 (3): 181–200.

Brunt, P. A. 1971. *Social Conflicts in the Roman Republic*. London: Chatto and Windus.

Buechler, Steven. 2011. *Understanding Social Movements: Theories from the Classical Era to the Present*. London: Routledge.

Bullard, Robert. 2005. *The Quest for Environmental Justice: Human Rights and the Politics of Pollution*. San Francisco: Sierra Club Books.

Bush, Ray. 2010. "Food Riots: Poverty, Power and Protest." *Journal of Agrarian Change* 10 (1): 119–129.

Cabrales, Sergio. 2019. *Terremoto Sociopolítico en Nicaragua: procesos, mecanismos y resultado de la oleada de protestas de 2018*. Pittsburgh: University of Pittsburgh. In press.

Calhoun, Craig. 1997. *Nationalism*. Minneapolis: University of Minnesota Press.

Calhoun, Craig. 2013. "Occupy Wall Street in Perspective." *British Journal of Sociology* 64 (1): 26–38.

Caniglia, Beth, Robert Brulle, and Andrew Szasz. 2015. "Civil Society, Social Movements, and Climate Change." In *Climate Change and Society*, ed. R. Dunlap and R. Brulle, 235–268. Oxford: Oxford University Press.

Caren, Neal, Sarah Gaby, and Catherine Herrold. 2017. "Economic Breakdown and Collective Action." *Social Problems* 64 (1): 133–155.

Caren, Neal, Kay Jowers, and Sarah Gaby. 2012. "A Social Movement Online Community: Stormfront and the White Nationalist Movement." *Research in Social Movements, Conflicts and Change* 33: 163–193.

Carley, Rob. 2019. *Culture and Tactics: Gramsci, Race, and the Politics of Practice*. Albany: State University of New York Press.

Carneiro, Robert L. 1970. "A Theory of the Origin of the State." *Science* 169 (August): 733–738.

Carneiro, Robert L. 1978. "Political Expansion as an Expression of the Principle of Competitive Exclusion." In *Origins of the State: The Anthropology of Political Evolution*, ed. Ronald Cohen and Elman R. Service, 205–223. Philadelphia: Institute for the Study of Human Issues.

Carroll, William K. 2007. "Hegemony and Counter-Hegemony in a Global Field." *Social Justice Studies* 1 (1): 36–67.

Carroll, William K. 2015a. "Modes of Cognitive Praxis in Transnational Alternative Policy Groups." *Globalizations* 12 (5): 710–727. doi:10.1080/147477 31.2014.1001231.

Carroll, William K. 2015b. "Robust radicalism." *Review of Radical Political Economics* 47 (4): 663–668.

Carroll, William K. 2016. *Expose, Oppose, Propose: Alternative Policy Groups and the Struggle for Global Justice.* New York: Zed Books.

Carroll, William K., and R. S. Ratner. 1996. "Master framing and cross-movement networking in contemporary social movements." *Sociological Quarterly* 37 (4): 601–625.

Carroll, William K., and J. P. Sapinski. 2013. "Embedding Post-Capitalist Alternatives? The Global Network of Alternative Knowledge Production and Mobilization." *Journal of World-Systems Research* 19 (2): 1–31.

Carroll, William K., and Jean Philippe Sapinski. 2017. "Transnational alternative policy groups in global civil society: Enablers of post-capitalist alternatives or carriers of NGOization?" *Critical Sociology* 43 (6): 875–892.

Castells, Manuel. 2012. *Networks of Outrage and Hope: Social Movements in yhe Internet Age.* Cambridge, UK: Polity Press.

Castells, Manuel. 2013. *Communication Power*, 2nd ed. Oxford: Oxford University Press.

Centeno, Manuel, and Joseph Cohen. 2010. *Global Capitalism: A Sociological Perspective.* London: Polity Press.

Centeno, Miguel, Manish Nag, Thayer S. Patterson, Andrew Shaver, and A. Jason Windawi. 2015. "The Emergence of Global Systemic Risk." *Annual Review of Sociology* 41.

Chase-Dunn, C. 1998. *Global Formation: Structures of the World-Economy.* Lanham, MD: Rowman & Littlefield.

Chase-Dunn, C. 1999. "Globalization: a world-systems perspective." *Journal of World-Systems Research* 5 (2): 187–215.

Chase-Dunn, C. 2005. "Global public Social Science." *The American Sociologist* 36 (3–4): 121–132. Reprinted in *Public Sociology: The Contemporary Debate*, ed. Lawrence T. Nichols, 179–194. New Brunswick, NJ: Transaction Press.

Chase-Dunn, C. 2006. "Globalization: A World-Systems Perspective." In *Global Social Change: Comparative and Historical Perspectives*, ed. C. Chase-Dunn, and S. Babones, 79–105. Baltimore: Johns Hopkins University.

Chase-Dunn, C. 2013. "BRICS and a Potentially Progressive Semiperiphery." In *BRICSs in Africa: Anti-Imperialist, Sub-Imperialist of in between?* ed. Patrick Bond, 56–58. South Africa: University of KwaZulu-Natal.

Chase-Dunn, C. 2014. "Continuities and Transformations in the Evolution of World-Systems" *Journal of Globalization Studies* 5 (1): 11–31.

Chase-Dunn, C., and E. Anderson, eds. 2005. *The Historical Evolution of World-Systems.* London: Palgrave.

Chase-Dunn, C., E. Anderson, H. Inoue, and A. Alvarez. 2015. "The Evolution of Economic Institutions: City-States and Forms of Imperialism since the

Bronze Age." IROWS Working Paper #79, https://irows.ucr.edu/papers/irows79/irows79.htm.

Chase-Dunn, C., and Ian Breckenridge-Jackson. 2014. "The Intermovement Network in the U.S. social forum process: Comparing Atlanta 2007 with Detroit 2010." IROWS Working Paper #71, http://irows.ucr.edu/papers/irows71/irows71.htm.

Chase-Dunn, C., James Fenelon, Thomas D. Hall, Ian Breckenridge-Jackson, and Joel Herrera. 2020. "Global Indigenism and the Web of Transnational Social Movements." In *New Frontiers of Globalization Research: Theories, Globalization Processes, and Perspectives from the Global South*, ed. Ino Rossi. New York: Springer.

Chase-Dunn, C., and Hiroko Inoue. 2012. "Accelerating Democratic Global State Formation." *Cooperation and Conflict* 47 (2): 157–175.

Chase-Dunn, C., and H. Inoue. 2017. "Problems of Peaceful Change: Interregnum, Deglobalization and the Evolution of Global Governance." Presented at the Annual Meeting of the International Studies Association, Baltimore. http//:irows.ucr.edu/papers/irows117/irows117.htm.

Chase-Dunn, C., and Hiroko Inoue. 2019. "A Multilevel Spiral model of Sociocultural Evolution: Polities and Interpolity Systems." Presented at the session on Theorizing Social Change at the Annual Meeting of the American Sociological Association, New York City. IROWS Working Paper #126, http://irows.ucr.edu/papers/irows126/irows126.htm.

Chase-Dunn, C., Hiroko Inoue, Teresa Neal, and Evan Heimlich. 2015. "The Development of World-Systems." *Sociology of Development* 1 (1).

Chase-Dunn, C., and Matheu Kaneshiro. 2009. "Stability and Change in the Contours of Alliances among Movements in the Social Forum Process." In *Engaging Social Justice*, ed. David Fasenfest, 119–133. Leiden: Brill.

Chase-Dunn, C., Yukio Kawano, and Benjamin Brewer. 2000. "Trade Globalization since 1795: Waves of Integration in the World-System." *American Sociological Review* 65 (February): 77–95.

Chase-Dunn, C., Roy Kwon, Kirk Lawrence, and Hiroko Inoue. 2011. "Last of the Hegemons: U.S. Decline and Global Governance." *International Review of Modern Sociology* 37 (1): 1–29.

Chase-Dunn, C., and B. Lerro. 2014. *Social Change: Globalization from the Stone Age to the Present*. London: Routledge.

Chase-Dunn, C., and Kelly M. Mann. 1998. *The Wintu and Their Neighbors*. Tucson: University of Arizona Press.

Chase-Dunn, C., Alessandro Morosin, and Alexis Álvarez. 2014. "Social Movements and Progressive Regimes in Latin America: World Revolutions and Semiperipheral Development." In *Handbook of Social Movements across*

Latin America, ed. Paul Almeida and Allen Cordero Ulate, 13–24. New York: Springer.

Chase-Dunn, C., and R. E. Niemeyer. 2009. "The World Revolution of 20xx." In *Transnational Political Spaces*, ed. Mathias Albert, Gesa Bluhm, Han Helmig, Andreas Leutzsch, and Jochen Walter, 35–57. Frankfurt/New York: Campus Verlag.

Chase-Dunn, C., Daniel Pasciuti, Alexis Alvarez, and Thomas D. Hall. 2006. "Waves of Globalization and Semiperipheral Development in the Ancient Mesopotamian and Egyptian World-Systems." In *Globalization and Global History*, ed. Barry Gills and William R. Thompson, 114–138. London: Routledge.

Chase-Dunn, C., Christine Petit, Richard Niemeyer, Robert A. Hanneman, and Ellen Reese. 2007. "The Contours of Solidarity and Division among Global Movements." *International Journal of Peace Studies* 12 (2): 1–15.

Chase-Dunn C., and Ellen Reese. 2007. "Global Party Formation in World Historical Perspective." In *Global Political Parties*, ed. Katarina Sehm-Patomaki and Marko Ulvila, 82–120. London: Zed Books.

Chase-Dunn, C., and Ellen Reese. 2008. "Global Party Formation in World Historical Perspective." In *Global Party Formation*, ed. Katarina Sehm-Patomaki and Marko Ulvila. London: Zed Books.

Chase-Dunn, C., E. Reese, M. Herkenrath, R. Giem, E. Gutierrez, L. Kim, and C. Petit. 2008. "North-South Contradictions and Bridges at the World Social Forum" In *North and South in the World Political Economy*, ed. R. Reuveny and W. R. Thompson, 341–366. Malden, MA: Blackwell.

Chen, Xi. 2012. *Social Protest and Contentious Authoritarianism in China*. Cambridge: Cambridge University Press.

Chesneaux, Jean, ed. 1972. *Popular Movements and Secret Societies in China, 1840–1950*. Palo Alto: Stanford University Press.

Chirot, Daniel. 1977. *Social Change in the 20th Century*. New York: Harcourt, Brace and Jovanovich.

Chun, J. J., G. Lipsitz, and Y. Shin. 2013. "Intersectionality as a Social Movement Strategy: Asian Immigrant Women Advocates." *Signs* 38 (4): 917–940.

Ciccariello-Maher, George. 2017. *Decolonizing Dialectics*. Durham, NC: Duke University Press.

Ciplet, David, J. Timmons Roberts, and Mizan R. Khan. 2015. *Power in a Warming World: The New Global Politics of Climate Change and the Remaking of Environmental Inequality*. Cambridge, MA: MIT Press.

Claudin, Fernando. 1975. *The Communist movement: From Comintern to Cominform*. London: Peregrine.

Cohn, Norman. 1970. *The Pursuit of the Millennium*. New York: Oxford University Press.

Cohn, Norman. 1993. *Cosmos, Chaos and the World to Come*. New Haven: Yale University Press.

Cohn, Samuel. 2012. "O'Connorian Models of Peripheral Development—or How Third World States Resist World Systemic Pressures by Cloning the Policies of States in the Core." In *Handbook of World Systems Analysis*, ed. Salvatore Babones and Christopher Chase-Dunn, 336–344. New York: Routledge.

Cohn, Samuel, and Rae Lesser Blumberg. 2016. "Introduction: crisis in development—how development lives and dies." In *Development in crisis: Threats to human well-being in the Global South and Global North*, ed. R. L. Blumberg and S. Cohn, 1–32. New York: Routledge.

Cohn, Samuel, and Gregory Hooks. 2016. "Introduction: A Manifesto for the Sociology of Development." In *The Sociology of Development Handbook*, ed. Gregory Hooks, 1–20. Berkeley: University of California Press.

Collins, John, and J. G. Manning. 2016. "Introduction." In *Revolt and Resistance in the Ancient Classical World and the Near East*, ed. J. Collins and J. G. Manning, 1–9. Leiden, Netherlands: Brill.

Collins, Patricia Hill. 2015. "Intersectionality's Definitional Dilemmas." *Annual Review of Sociology* 41: 1–20.

Collins, Patricia Hill, and S. Bilge. 2016. *Intersectionality*. Malden, MA: Polity Press.

Comintern Electronic Archive. n.d. http://www.comintern-online.com/.

Cordero Ulate, Allen. 2019. *Resumen Lucha Contra el Combo Fiscal (10 de setiembre-10 de diciembre del 2018)*. Unpublished Manuscript. Department of Sociology, University of Costa Rica.

Coyne, Gary, Juliann Allison, Ellen Reese, Katja Guenther, Ian Breckenridge-Jackson, Edwin Elias, Ali Lairy, James Love, Anthony Roberts, Natasha Rodojcic, Miryam Ruvalcaba, Elizabeth Schwarz, and Christopher Chase-Dunn. 2010. "2010 U.S. Social Forum Survey of Attendees: Preliminary Report." IROWS Working Paper #64, http://irows.ucr.edu/papers/irows64/irows64.htm.

Crenshaw, Kimberlé. 2020. *On Intersectionality: Essential Writings*. New York: The New Press.

Crouch, Colin, and Wolfgang Streeck, eds. 1997. *Political Economy of Modern Capitalism: Mapping Convergence and Diversity*. London: Sage.

Curran, Michaela, Elizabeth A. G. Schwarz, and C. Chase-Dunn. 2014. "The Occupy Movement in California." In *What Comes After Occupy?: The Regional Politics of Resistance*, ed. Todd A. Comer. Cambridge Scholars Publishing, http://irows.ucr.edu/papers/irows74/irows74.htm.

Dale, John G. 2011. *Free Burma: Transnational Legal Action and Corporate Accountability*. Minneapolis: University of Minnesota Press.

Davis, Jeffrey E. 2010. *Hand Talk: Sign Language among American Indian Nations.* Cambridge: Cambridge University Press.

Davis, Mike. 1992. *City of Quartz: Excavating the History of Los Angeles.* London: Verso.

Davis, Mike. 2006. *Planet of Slums.* London: Verso.

Davis, Mike. 2018. *Old Gods, New Enigmas: Marx's Lost Theory.* London: Verso.

della Porta, Donatella. 2005. "Multiple Belongings, Tolerant Identities, and the Construction of 'Another Politics': Between the European Social Forum and the Local Social Fora." In *Transnational Protest and Global Activism*, ed. by Donatella della Porta and Sidney Tarrow, 175–202. Lanham, MD: Rowman & Littlefield.

della Porta, Donatella. 2015. *Social Movements in Times of Austerity: Bringing Capitalism back into Protest Analysis.* London: Polity Press.

della Porta, Donatella, Joseba Fernández, Hara Kouki, and Lorenzo Mosca. 2017. *Movement parties against austerity.* Malden. MA: Polity Press.

de Moor, Joost, Katrin Uba, Mattias Wahlstrom, Magnus Wennerhag, Michiel De Vydt, Paul Almeida, et al. 2020. "Introduction: Fridays For Future—An Expanding Climate Movement." In *Protest for a Future II: Composition, Mobilization, and Motives of the Participants in Fridays For Future Climate Protests on 20–27 September, 2019, in 19 Cities around the World*, ed. Joost de Moor, Katrin Uba, Mattias Wahlström, Magnus Wennerhag, and Michiel De Vydt, 6-33. doi:10.17605/OSF.IO/ASRUW.

Denmark, Robert. 2008. "Fundamentalisms as Global Social Movements." *Globalizations* 5 (4): 571–582.

Dennis, Brady, and Chris Mooney. 2018. "'We Are in Trouble.' Global Carbon Emissions Reached a New Record High in 2018." *Washington Post*, December 5.

Diani, Mario, and Maria Kousis. 2014. "The duality of claims and events: the Greek campaign against the Troika's memoranda and austerity, 2010–2012." *Mobilization* 19 (4): 387–404.

Dicken, Peter. 1986. *Global Shift: Industrial Change in a Turbulent World.* New York: Harper and Row.

Dicken, Peter. 2015. *Global Shift: Mapping the Changing Contours of the World Economy*, 7th ed. New York: Guilford.

DiEM25. 2015. "Democracy in Europe Movement." https://diem25.org/.

Dixon, Marc, and Andrew Martin. 2012. "We Can't Win This on Our Own: Unions, Firms, and the Mobilization of External Allies in Labor Disputes." *American Sociological Review* 77: 946–969.

Dreiling, Michael, and Derek Darves. 2016. *Agents of Neoliberal Globalization: Corporate Networks, State Structures, and Trade Policy*. Cambridge: Cambridge University Press.

DuBois, Cora. (1939) 2007. *The 1870 Ghost Dance*. Lincoln: University of Nebraska Press.

Dunbar-Ortiz, Roxanne. 2014. *An Indigenous Peoples' History of the United States*. Boston: Beacon Press

Durkheim, Emile. 1915. *The Elementary Forms of Religious Life*. New York: Macmillan.

Dwyer, Peter, and Louis Zeilig. 2012. *African Struggles Today: Social Movements since Independence*. Chicago: Haymarket Books.

Eckstein, Susan E. 2002. "Globalization and mobilization: resistance to neoliberalism in Latin America." In *The New Economic Sociology: Developments in an Emerging Field*, ed. M. Guillén, R. Collins, P. England, and M. Meyer, 330–368. New York: Russell Sage.

Edelman, Marc. 1999. *Peasants against Globalization*. Stanford, CA: Stanford University Press.

Edwards, Bob, and John D. McCarthy. 2004. "Resources and Social Movement Mobilization." In *The Blackwell Companion to Social Movements*, ed. D. Snow, S. Soule, and H. Kriesi, 116–152. Oxford: Blackwell.

Edwards, Bob, John D. McCarthy, and Dane Mataic. 2018. "The Resource Context of Social Movements." In *The Wiley-Blackwell Companion to Social Movements*, ed. D. Snow, S. Soule, H. Kriesi, and H. McCammon, 79–97. Oxford: Blackwell.

Eichstedt, Jennifer L. 2001. "Problematic White Identities and a Search for Racial Justice." *Sociological Forum* 16 (3): 445–470.

Eisenstadt, S. N. 1986. *The Origins and Diversity of Axial Age Civilizations*. Albany: State University of New York Press.

Ellacuría, Ignacio. 1985. *Conversión de la Iglesia al Reino de Dios: para anunciarlo y realizarlo en la historia*. San Salvador: UCA Editores.

Emerson, Thomas E., and Kristin M. Hedman. 2016. "The Dangers of Diversity: The Consolidation and Dissolution of Cahokia, Native North America's First Urban Polity." In *Beyond Collapse: Archaeological Perspectives on Resilience, Revitalization, and Transformation in Complex Societies*, ed. Ronald K. Faulseit, 147–175. Carbondale, IL: Southern Illinois University Press.

Eschle, Catherine, and Neil Stammers. 2004. "Taking Part: Social Movements, INGOs, and Global Change." *Alternatives: Global, Local, Political* 29: 333–372.

Evans, Peter B. 2009. "From Situations of Dependency to Globalized Social Democracy." *Studies in Comparative International Development* 44: 318–336

Evans, Peter B. 2010. "Is it Labor's Turn to Globalize? Twenty-first Century Opportunities and Strategic Responses." *Global Labour Journal* 3 (1): 352–379.

Evans, Peter B. 2020. "Transnational Social Movements" In *The New Handbook of Political Sociology*, ed. Thomas Janoski, Cedric De Leon, Joya Misra, and Isaac William Martin. *The New Handbook of Political Sociology*, 1053–1077. Cambridge: Cambridge University Press.

Evans, Peter B., and James E. Rauch. 1999. "Bureaucracy and Growth: A Cross-National Analysis of the Effects of 'Weberian' State Structures on Economic Growth." *American Sociological Review* 64 (5): 748–765.

Extinction Rebellion. 2019. *This is not a Drill: An Extinction Rebellion Handbook*. New York: Penguin Books.

Fagan, Brian M. 1991. *Ancient North America: The Archaeology of a Continent*. London: Thames and Hudson.

Fenelon, James. 1998. *Culturicide, Resistance and Survival of the Lakota*. New York: Garland Publishing.

Fisher, Dana. 2007. "Taking Cover Beneath the Anti-Bush Umbrella: Cycles of Protest and Movement-to-Movement Transmission in an Era of Repressive Politics." *Research in Political Sociology* 15: 27–56.

Fisher, William F., and Thomas Ponniah, eds. 2003. *Another World is Possible: Popular Alternatives to Globalization at the World Social Forum*. London: Zed Books.

Flacks, Richard. 1988. *Making History: The American Left and the American Mind*. New York: Columbia University Press.

Flannery, Kent, and Joyce Marcus. 2012. *The Creation of Inequality*. Cambridge, MA: Harvard University Press.

Flores, Edward Orozco. 2018. *"Jesus Saved an Ex-Con": Political Activism and Redemption after Incarceration*. New York: New York University Press.

Foran, John. 2005. *Taking Power: On the Origins of Third World Revolutions*. Cambridge: Cambridge University Press.

Foran, John. 2014. *"Get it Done!" The Global Climate Justice Movement's Struggle to Achieve a Radical Climate Treaty*. Unpublished manuscript. University of California, Santa Barbara, Dept. of Sociology.

Foran, John. 2018. *Taking or (re)Making Power?: The New Movements for Radical Social Change and Global Justice*. London: Zed Books.

Foster, John Bellamy, Brett Clark, and Richard York. 2011. *The Ecological Rift: Capitalism's War On The Earth*. New York: New York University Press.

Fox, Cybelle. 2012. *Three Worlds of Relief: Race, Immigration, and the American Welfare State from the Progressive Era to the New Deal*. Princeton: Princeton University Press.

Frank, Dana. 2018. *The Long Honduran Night: Resistance, Terror, and the United States in the Aftermath of the Coup*. Chicago: Haymarket Books.

Freeden, Michael. 2003. *Ideology*. New York: Oxford University Press.

Freeman, Jo. 1973. "The Tyranny of Structuralessness." *Berkeley Journal of Sociology* 17: 151–165.

Fukuyama, Francis. 2013. "The Middle-Class Revolution." *Wall Street Journal*, June 29. http://online.wsj.com/article/SB10001424127887323873904578 571472700348086.html?KEYWORDS=francis+fukuyama.

Gallo-Cruz, Selina. 2019. "Nonviolence beyond the state: International NGOs and local nonviolent mobilization." International Sociology 34(6): 655–674.

Gamble, Lynn. 2008. *The Chumash at European Contact*. Berkeley: University of California Press.

Ganz, Marshall. 2009. *Why David Sometimes Wins: Leadership, Organization, and Strategy in the California Farm Worker Movement*. Oxford: Oxford University Press.

Garland. David. 2016. *The Welfare State: A Very Short Introduction*. Oxford: Oxford University Press.

Garrelts, Heiko, and Matthias Dietz. 2014. "Introduction: contours of the transnational climate movement conception and contents of the handbook." In *Routledge Handbook of the Climate Change Movement*, 1–16. New York: Routledge.

Gayton, Anna H. 1930. "The Ghost Dance of South Central California." *American Archaeology and Ethnology* 28 (2): 57–88.

Gest, Justin. 2016. *The New Minority: White Working Class Politics in an Age of Immigration and Inequality*. Oxford: Oxford University Press.

Gill, Stephen. 2000. "Toward a Post-Modern Prince?: The Battle of Seattle as a Moment in the New Politics of Globalization." *Millennium* 29 (1): 131–140.

Gill, Stephen. 2003. *Power and Resistance in the New World Order*. London: Palgrave.

Gillan, Kevin. 2018. "Temporality in social movement theory: vectors and events in the neoliberal timescape." *Social Movement Studies*: 1–21. doi:10.1080 /14742837.2018.1548965.

Gills, Barry, and Christopher Chase-Dunn. 2019. "In search of unity: a new politics of solidarity and action for confronting the crisis of global capitalism" *Globalizations* 16 (7): 967–972. doi:10.1080/14747731.2019.1655889.

Gitlin, Todd. 1993. *The Sixties: Years of Hope, Days of Rage*. New York: Bantam Books.

Gitlin, Todd. 2012. *Occupy Nation*. New York: HarperCollins.

Goldfrank, W. L. 1978. "Fascism and World Economy." In *Social Change in the Capitalist World Economy*, ed. Barbara Hockey Kaplan, 75–120. Beverly Hills, CA: Sage.

Goldstein, Joshua. 1988. *Long Cycles: Prosperity and War in the Modern Age*. New Haven: Yale University Press.

Goldstone, Jack A. 1991. *Revolution and Rebellion in the Early Modern World*. Berkeley: University of California Press.

Goldstone, Jack A. 2014. *Revolutions*. New York: Oxford University Press.

Goldstone, Jack A., and Charles Tilly. 2001. "Threat (and Opportunity): Popular Action and State Response in the Dynamic of Contentious Action." In *Silence and Voice in the Study of Contentious Politics*, ed. R. Aminzade, J. Goldstone, D. McAdam, E. Perry, W. Sewell, S. Tarrow, and C. Tilly, 179–194. Cambridge: Cambridge University Press.

Goodwin, Jeff. 2001. *No Other Way Out: States and Revolutionary Movements, 1945–1991*. Cambridge: Cambridge University Press.

Gordon, L. 2017. *The Second Coming of the KKK*. New York: Norton.

Gottlieb, Robert. 1993. *Forcing the Spring: The Transformation of the American Environmental Movement*. New York: Island Press.

Gould, Jeffrey, and Aldo Lauria-Santiago. 2008. *To Rise in Darkness: Revolution, Repression, and Memory in El Salvador, 1920–1932*. Durham, NC: Duke University Press.

Gould, Kenneth A., David N. Pellow, and Allan Schnaiberg. 2004. "Interrogating the Treadmill of Production: Everything You Wanted to Know About the Treadmill but Were Afraid to Ask." *Organization & Environment* 17 (3): 296–316.

Gould, Roger V. 1995. *Insurgent Identities: Class, Community, and Protest in Paris from 1848 to the Commune*. University of Chicago Press.

Graeber, David. 2013. *The Democracy Project: A History, A Crisis, A Movement*. New York: Spiegel and Grau (Random House).

Graham, Helen, and Paul Preston, eds. 1987. *The Popular Front in Europe*. New York: St. Martin's Press.

Gramsci, Antonio. 1971. *Selections from the Prison Notebooks*. New York: International Publishers. http://courses.justice.eku.edu/pls330_louis/docs/gramsci-prison-notebooks-vol1.pdf.

Grimes, Peter. 1999. "The Horsemen and the Killing Fields: The Final Contradiction of Capitalism." In *Ecology and the World-System*, ed. Walter Goldfrank, David Goodman, and Andrew Szasz, chapter 2. Westport, CT: Greenwood Press.

Grimes, Peter. 2003. "Sociologist of Energy: Howard Ehrlich interviews Peter Grimes." *Social Anarchism: A Journal of Theory and Practice* 34: 37–50.

Hadden, Jennifer. 2014. "Explaining Variation in Transnational Climate Change Activism: The Role of Inter-Movement Spillover." *Global Environmental Politics* 14 (2): 7–25

Hadden, Jennifer. 2015. *Networks in Contention: The Divisive Politics of Climate Change.* Cambridge: Cambridge University Press.

Haglund, La Dawn. 2010. *Limiting Resources: Market-Led Reform and the Transformation of Public Goods.* University Park: Pennsylvania State University Press.

Hall, Thomas D., and J. V. Fenelon. 2009. *Indigenous Peoples and Globalization: Resistance and Revitalization.* Boulder, CO: Paradigm Press.

Halpern, Abraham M. 1988. "Southeastern Pomo Ceremonials: The Kuksu Cult and Its Successors." In *University of California Publications, Anthropological Records*, vol. 29. Berkeley: University of California Press.

Harari, Yuval Noah. 2014. *Sapiens: A Brief History of Humankind.* New York: Random House.

Hardt, Michael, and Antonio Negri. 2012. *Declaration.* New York: Argo Navis Author Services.

Harlan, Sharon, David Pellow, J. Timmons Roberts, and Shannon Elizabeth Bell. 2015. "Climate Justice and Inequality." In *Climate Change and Society*, ed. R. Dunlap and R. Brulle, 93–126. Oxford: Oxford University Press.

Harris, Jerry, Carl Davidson, Bill Fletcher, and Paul Harris. 2017. "Trump and American Fascism." *International Critical Thought* 7 (4): 476–492. doi:10.1 080/21598282.2017.1357491.

Harris, Kevan. 2012. "The Brokered Exuberance of the Middle Class: An Ethnographic Analysis of Iran's 2009 Green Movement." *Mobilization* 17 (4): 435–455.

Harris, Kevan, and Ben Scully. 2015. "A hidden counter-movement? Precarity, politics, and social protection before and beyond the neoliberal era." *Theory and Society* 44: 415–444.

Harris, Marvin. 1977. *Cannibals and Kings.* New York: Random House.

Harvey, David. 2005. *A Brief History of Neoliberalism.* Oxford: Oxford University Press.

Hendrix, Cullen, and Stephan Haggard. 2015. "Global food prices, regime type, and urban unrest in the developing world." *Journal of Peace Research* 52 (2): 143–157.

Herkenrath, Mark. 2011. *Die Globalisierung der Sozialen Bewegungen: Transnationale Zivilgesellschaft und die Suche nach einer gerechten Weltordnung.* Wiesbaden: Springer.

Herrera, Joel S. 2014. "Neoliberal Reform and Latin America's Turn to the Left." Honors thesis, University of California-Riverside. Submitted for publication.

Herrmann, Edward W., G. William Monaghan, William F. Romain, Timothy M. Schilling, Jarrod Burks, Karen L. Leone, Matthew P. Purtill, and Alan C. Tonetti. 2014. "A New Multistage Construction Chronology for the Great Serpent Mound, USA." *Journal of Archaeological Science* 50: 117–125. http://www.sciencedirect.com/science/article/pii/S030544031400246.

Hignett, Katherine. 2019. "Ancient Mesopotamia: ritual child sacrifice uncovered in Bronze Age Turkey." *Newsweek*, August 13. https://www.newsweek.com/ancient-mesopotamia-child-sacrifice-turkey-archaeology-1000934.

Hill, John E. 2015. *Through the Jade Gate to Rome: A Study of the Silk Routes, 1st to 2nd centuries CE. An Annotated translation from the Hou Hanshu "The Chronicle on the Western Regions"*, vol. 1. CreateSpace Independent Publishing Platform.

Hobsbawm, Eric J. 1957. *Primitive Rebels: Studies in Archaic Forms of Social Movement in the 19th and 20th Centuries*. England: Manchester University Press.

Hobsbawm, Eric J. 1994. *The Age of Extremes: A History of the World, 1914–1991*. New York: Pantheon.

Hochschild, Adam. 2005. *Bury the Chains: Prophets and Rebels in the Fight to Free an Empire's Slaves*. New York: Houghton Mifflin.

Hochschild, Arlie R. 2016. *Strangers in Their Own Land*. New York: New Press.

Holloway, John. 2002. *Change the World Without Taking Power: The Meaning of Revolution Today*. London: Pluto Press.

Honari, Ali. 2019. "Online and Offline Political Participation Under Repression: Iranian Green Movement Supporters between Two Elections, 2009–2013." PhD Diss., Department of Sociology, Free University of Amsterdam.

Huber, Evelyne, and John D. Stephens. 2012. *Democracy and the left: social policy and inequality in Latin America*. Chicago: University of Chicago Press.

Hudis, Peter, and Kevin Anderson, eds. 2004. *The Rosa Luxemburg Reader*. New York: Monthly Review Press.

Hudson, Travis. 1982. *Guide to Painted Cave*. Santa Barbara, CA: McNally and Loftin Hudson.

Hung, Ho-Fung. 2011. *Protest with Chinese Characteristics: Demonstrations, Riots and Petitions in the Mid-Qing Dynasty*. New York: Columbia University Press.

Hunt, Lynn. 2007. *Inventing Human Rights*. New York: Norton.

Hutter, S., H. Kriesi, and J. Lorenzini. 2018. "Social Movements in Interaction with Political Parties." In *The Wiley-Blackwell Companion to Social Move-*

ments, 2nd ed., ed. D. A. Snow, S. A. Soule, H. Kriesi, and H. J. McCammon, 322–337. Oxford: Blackwell.

Inclán, María. 2008. "From the ¡Ya Basta! to the Caracoles: Zapatista Mobilization under Transitional Conditions." *American Journal of Sociology* 113 (5): 1316–1350.

Inclán, María. 2018. *The Zapatista Movement and Mexico's Democratic Transition: Mobilization, Success, and Survival.* Oxford: Oxford University Press.

Inoue, Hiroko, Alexis Álvarez, E. N. Anderson, Kirk Lawrence, Teresa Neal, Dmytro Khutkyy, Sandor Nagy, Walter De Winter, and Christopher Chase-Dunn. 2016. "Comparing World-Systems: Empire Upsweeps and Non-core marcher states Since the Bronze Age," IROWS Working Paper #56.

Inoue, Hiroko, and C. Chase-Dunn. 2018. "Spirals of Sociocultural Evolution within Polities and in Interpolity Systems." Presented at the annual meeting of the American Sociological Association, Philadelphia, PA, August.

James, C. L. R. 1963. *The Black Jacobins: Toussaint L'Ouverture and the San Domingo Revolution.* New York: Random House.

James, C. L. R. 2018. *Revolutionary Marxism: Selected Writings 1939–1949.* Chicago: Haymarket Books

James, Paul, and Manfred Steger. 2014. "A Genealogy of 'Globalization': The Career of a Concept." *Globalizations* 11 (4): 417–434.

Jasper, James. 2018. *The Emotions of Protest.* Chicago: University of Chicago Press.

Jenkins, J. Craig, and Kurt Schock. 2004. "Political process, international dependence, and mass political conflict: A global analysis of protest and rebellion, 1973–1978." *International Journal of Sociology,* 33 (4): 41–63.

Jennings, Justin. 2010. *Globalizations and the Ancient World.* Cambridge: Cambridge University Press.

Joas, Hans. 2013. *The Sacredness of the Person: A New Genealogy of Human Rights.* Washington, DC: University of Georgetown Press.

Johnson, Erik W., and Scott Frickel. 2011. "Ecological threat and the founding of US national environmental movement organizations, 1962–1998." *Social Problems* 58 (3): 305–329.

Johnson, John. n.d. "California Indian Religions." http://www.anth.ucsb.edu/classes/anth131ca/CA%20Indian%20Religion.pdf.

Johnston, Hank, and Paul Almeida, eds. 2006. *Latin American Social Movements: Globalization, Democratization and Transnational Networks.* Lanham, MD: Rowman & Littlefield.

Jorgenson, Andrew. 2012. "The Sociology of Ecologically Unequal Exchange and Carbon Dioxide Emissions, 1960–2005." *Social Science Research* 41 (2): 242–252.

Juergensmeyer, Mark. 2003. *Terror in the Mind of God*. Berkeley: University of California Press.

Juris, Jeffrey. 2008. *Networking Futures: The Movements Against Corporate Globalization*. Durham: Duke University Press.

Kaldor, Mary. 2003. *Global Civil Society*. Malden, MA: Polity Press.

Kalleberg, Arne L. 2009. "Precarious Work, Insecure Workers: Employment Relations in Transition" *American Sociological Review*. 74 (1): 1–22.

Kalleberg, Arne L. 2011. *Good Jobs, Bad Jobs: The Rise of Polarized and Precarious Employment*. New York: Russell Sage.

Kanellopoulos, K., K. Kostopoulos, D. Papanikolopoulos, and R. Vasileios. 2017. "Competing modes of coordination in the Greek anti-austerity campaign, 2010–2012." *Social Movement Studies* 16 (1): 101–118

Kaneshiro, Matheu, Kirk S. Lawrence, and C. Chase-Dunn. 2015. "Global environmentalists and their movements at the World Social Forums" In *Handbook on World Social Forum Activism*, ed. Jackie Smith, Scott Byrd, Ellen Reese, and Elizabeth Smythe, 186–205. Boulder, CO: Paradigm Publishers.

Karatasli, Savan Savas. 2018. "Global Waves of Secessionism in the West, 1492–Present." In *Return of Geopolitics*, ed. Albert Bergesen and Christian Suter, 69–96. New York: LitVerlag.

Karatasli, Savan Savas, Sefika Kumral, Ben Scully, and Smriti Upadhyay. 2014. "Class, Crisis, and the 2011 Protest Wave: Cyclical and Secular Trends in Global Labor Unrest." In *Overcoming Global Inequalities*, ed. Immanuel Wallerstein, Christopher Chase-Dunn, and Christian Suter. Boulder, CO: Paradigm Publishers.

Karides, Marina, Walda Katz-Fishman, Rose M. Brewer, Jerome Scott, and Alice Lovelace. 2010. *The United States Social Forum: Perspectives of a Movement*. Chicago: Changemaker.

Kay, Cristóbal. 1989. *Latin American Theories of Development and Underdevelopment*. New York: Routledge.

Kay, Tamara, and Rhonda Lynn Evans. 2018. *Trade Battles: Activism and the Politicization of International Trade Policy*. Oxford: Oxford University Press.

Keck, Margaret, and Katherine Sikkink. 1998. *Activists Beyond Borders: Advocacy Networks in International Politics*. Ithaca: Cornell University Press.

Kentikelenis, Alexander E., and Sarah Babb. 2019. "The Making of Neoliberal Globalization: Norm Substitution and the Politics of Clandestine Institutional Change." *American Journal of Sociology* 124 (6): 1720–1762.

Kenworthy, Lane. 2019. *Social Democratic Capitalism*. New York: Oxford University Press.

Klandermans, Bert. 1997. *The Social Psychology of Protest*. Oxford: Blackwell.

Klein, Richard G., and Blake Edgar. 2002. *The Dawn of Human Culture*. New York: John Wiley.

Kohli, Atul. 2004. *State-Directed Development: Political Power and Industrialization in the Global Periphery*. Cambridge: Cambridge University Press.

Korotayev, A. V., A. Malkov, and D. Khaltourina. 2006. *Introduction to Social Macrodynamics: Secular Cycles and Millennial Trends*. Moscow: Editorial URSS.

Korotayev, A. V., and J. V. Zinkina. 2011. "Egyptian Revolution: A Demographic Structural Analysis." *Entelequia, Revista Interdisciplinar* 13: 139–170.

Kousis, Maria. 2015. "The Transnational Dimension of the Greek Protest Campaign against Troika Memoranda and Austerity Policies, 2010–2012." In *Spreading Protests: Social Movements in Times of Crisis*, ed. D. della Porta and A Mattoni, 139–170. Colchester, UK: ECPR Press.

Kousis, Maria. 2016. "The spatial dimensions of the Greek protest campaign against the Troika's memoranda and austerity, 2010–2013." In *Street Politics in the Age of Austerity: From the Indignados to Occupy*, ed. M Ancelovici, P. Dufour, and H. Nez, 147–174. Netherlands: Amsterdam University Press.

Kriesi, Hans Peter, Edgar Grande Romain Lachat, Martine Dolezal, Simon Bornschier, and Timitheos Frey. 2006. "Globalization and the Transformation of the National Political Space: Six European Countries Compared." *European Journal of Political Research* 45, 921–956.

Kriesi, Hans, Edgar Grande Romain Lachat, Martine Dolezal, Simon Bornschier, and Timitheos Frey. 2008. *West European Politics in the Age of Globalization*. Cambridge: Cambridge University Press.

Krinsky, John, and Ellen Reese. 2006. "Forging and Sustaining Labor-Community Coalitions: The Workfare Justice Movement in Three Cities." *Sociological Forum* 21 (4): 623–658.

Krishna, A. 2002. *Active Social Capital: Tracing the Roots of Development and Democracy*. New York: Columbia University Press.

Kuecker, Glen. 2007. "The Perfect Storm." *International Journal of Environmental, Cultural and Social Sustainability* 3 (5): 1–10. https://scholarship.depauw.edu/hist_facpubs/3/.

Kumral, Sefika. 2015. "Hegemonic Transition, War and Opportunities for Fascist Militarism" In *The Longue Durée of the Far Right*, ed. Richard Saull et al. London: Routledge.

Laqueur, Walter Z., ed. 1976. *Fascism: A Reader's Guide*. Berkeley: University of California Press.

Laqueur, Walter Z., and George L. Mosse, eds. 1966. *International Fascism, 1920–1945*. New York: Harper and Row.

Lawrence, Peter. 1964. *Road Belong Cargo: A Study of the Cargo Movement in the Southern Madang District*. New Guinea Manchester: Manchester University Press.

Leal, Diego. 2020. "Mass Transit Shutdowns as a Tactical Innovation in Bogotá, Colombia." *Social Currents* (published online February 21). doi:10.1177/2329496520906826.

Ledeen, M. A. 1972. *Universal Fascism: They Theory and Practice of the Fascist International, 1928–1936*. New York: Howard Fertig.

Lekson, Stephen H. 1999. *The Chaco Meridian: Centers of Political Power in the Ancient Southwest*. Walnut Creek, CA: Altamira Press.

Lee, Ching Kwan. 2007. *Against the Law: Labor Protests in China's Rustbelt and Sunbelt*. Berkeley: University California Press.

Lee, Harry F. 2018. "Internal Wars in History: Triggered by Natural Disasters or Socio-Ecological Catastrophes?" *The Holocene* 28 (7): 1071–1081. doi:10.1177/0959683618761549.

Lee, J. S. 1931. "The Periodic Recurrence of Internecine Wars in China." *China Journal of Science and Arts* 14: 114.

Lee, J. S. 1933. "The Periodic Recurrence of Internecine Wars in China." Studies presented to Ts'ai Yuan p'ei on his sixty-fifth birthday, ed. Fellows and Assistants of the Institute of History and Philology), vol. 1, 157–166. Taipei: Institute of History and Philology.

Levitsky, Steven, and Kenneth Roberts. 2011. "Introduction. Latin America's 'Left Turn': A Framework for Analysis." In *The Resurgence of the Latin American Left,* ed. Steven Levitsky and Kenneth M. Roberts, 1–30. Baltimore: Johns Hopkins University Press.

Lichbach, Mark Irving. 2003. "The Anti-Globalization Movement: A New Kind of Protest." In *Peace and Conflict*, ed. Monty G. Marshall and Ted Robert Gurr, 39–42. College Park, MD: Center for International Development and Conflict Management, University of Maryland.

Lieberman, Victor. 2003. *Strange Parallels: Southeast Asia in Global Context, c. 800–1830. Vol. 1: Integration on the Mainland*. Cambridge: Cambridge University Press.

Lieberman, Victor. 2009. *Strange Parallels: Southeast Asia in Global Context, c. 800–1830. Vol 2: Mainland Mirrors: Europe, Japan, China, South Asia, and the Islands*. Cambridge: Cambridge University Press.

Lindh, A. 2015. "Public opinion against markets? Attitudes towards market distribution of social services—a comparison of 17 countries." *Social Policy Administration* 49 (7): 887–910.

Lindholm, Charles, and Jose Pedro Zuquete. 2010. *The Struggle for the World: Liberation Movements for the 21st Century*. Palo Alto: Stanford University Press.

Linz, Juan J. 1976. "Some Notes on a Comparative Study of Fascism in Sociological Historical Perspective." In *Fascism: A Reader's Guide*, ed. Walter Z., Laqueur, 3–121. Berkeley: University of California Press.

Lobao, Linda. 2016. "The sociology of subnational development: conceptual and empirical foundations." In *The Sociology of Development Handbook*, ed. G. Hooks, P. Almeida, D. Brown, S. Cohn, S. Curran, et al., 265–292. Berkeley: University of California Press.

Lu, Y., and R. Tao. 2017. "Organizational Structure and Collective Action: Lineage Networks, Semiautonomous Civic Associations, and Collective Resistance in Rural China." *American Journal of Sociology* 122 (6): 1726–1774.

Luft, Rachel E. 2009. "Beyond Disaster Exceptionalism: Social Movement developments in New Orleans after Hurricane Katrina." *American Quarterly* 61 (3): 499–527.

Luna, Zakiya T. 2016. "'Truly a Women of Color Organization': Negotiating Sameness and Difference in Pursuit of Intersectionality." *Gender and Society* 30 (5): 769–790.

Lynd, Staughton, and Andrej Grubacic. 2008. *Wobblies and Zapatistas: Conversations on Anarchism, Marxism and Radical History*. Oakland, CA: PM Press.

MacMullen, Ramsay. 1963. "A Note on Roman Strikes." *Classical Journal* 58 (6): 269–271. https://www.jstor.org/stable/3293990.

Mahoney, James. 2010. *Colonialism and Postcolonial Development: Spanish America in Comparative Perspective*. Cambridge: Cambridge University Press.

Mann, Charles C. 2005. *1491: New Revelations of the Americas Before Columbus*. New York: Vintage.

Mann, Michael. 2004. *Fascists*. Cambridge: Cambridge University Press.

Mann, Michael. 2013. *The Sources of Social Power, vol. 4: Globalizations, 1945–2011*. New York: Cambridge University Press.

Markoff, John. 1996. *Waves of Democracy: Social Movements and Political Change*. Thousand Oaks, CA: Pine Forge Press.

Markoff, John. 2015a. "Historical Analysis and Social Movements Research." In *The Oxford Handbook of Social Movements*, ed. D. Della Porta and M. Diani. 68–85. Oxford: Oxford University Press.

Markoff, John. 2015b. *Waves of Democracy: Social Movements and Political Change*, 2nd ed. Thousand Oaks, CA: Sage.

Marshall, M. G., and K. Jaggers. 2009. *Polity IV Project: Dataset Users' Manual*. Vienna, VA: Center for Systemic Peace.

Marshall, T. H. 1950. *Citizenship and Social Class: And Other Essays.* Cambridge: Cambridge University Press.

Martin, William G., ed. 2008. *Making Waves: Worldwide Social Movements, 1750–2005.* Boulder, CO: Paradigm.

Marx, Karl. 1844. *A Contribution to the Critique of Hegel's Philosophy of Right.* http://www.marxists.org/archive/marx/works/1843/critique-hpr/intro.htm.

Marx, Karl. 1852. *The 18th Brumaire of Napoleon Bonaparte.* New York: International Publishers.

Marx, Karl. 1972. *The Eighteenth Brumaire of Louis Bonaparte.* New York: International Publishers.

Mason, Paul. 2013. *Why It's Still Kicking Off Everywhere: The New Global Revolutions.* London: Verso.

Mason, Paul. 2015. *Postcapitalism: A Guide to Our Future.* New York: Farrar, Straus and Giroux.

Massey, Douglas. 2005. *Strangers in a Strange Land: Humans in an Urbanizing World.* New York: WW Norton.

McAdam, Doug. 2003. "Beyond Structural Analysis: Toward a More Dynamic Understanding of Social Movements." In *Social Movement Analysis: The Network Perspective,* ed. Mario Diani and Doug McAdam, 281–299. Oxford: Oxford University Press.

McAdam, Doug, and Hilary Boudet. 2012. *Putting Social Movements in Their Place: Explaining Opposition to Energy Projects in the United States, 2000–2005.* New York: Cambridge University Press.

McCammon, Holly, and Nella Van Dyke. 2010. "Applying Qualitative Comparative Analysis to Empirical Studies of Social Movement Coalition Formation." In *Strategic Alliances: Coalition Building and Social Movements,* ed. N. Van Dyke and H. McCammon, 292–315. Minneapolis: University of Minnesota Press.

McCarthy, John D., and Mayer N. Zald. 1977. "Resource Mobilization and Social Movements: A Partial Theory." *American Journal of Sociology* 82 (6): 1212–1241.

McIlwaine, Cathy. 2007. "From Local to Global to Transnational Civil Society: Reframing Development Perspectives on the Non-State Sector." *Geography Compass* 1 (6): 1252–1281.

McIntosh, Peggy. 2007. "White Privilege and Male Privilege." In *Race, Ethnicity, and Gender: Selected Readings,* ed. Joseph F. Healey and Eileen O'Brien, 377–384. Los Angeles: Pine Forge Press.

McMichael, Philip. 2016. *Development and Social Change: A Global Perspective,* 6th ed. Thousand Oaks, CA.: Pine Forge.

McVeigh, Rory, and Kevin Estep. 2019. *The Politics of Losing: Trump, the Klan, and the Mainstreaming of Resentment.* New York: Columbia University Press.

Mesa-Lago, Carmelo. 2007. "The Extension of Healthcare Coverage and Protection in Relation to the Labour Market: Problems and Policies in Latin America." *International Social Security Review* 60 (1): 3–31.

Mesa-Lago, Carmelo. 2008. *Reassembling Social Security: A Survey of Pensions and Health Care Reforms in Latin America.* Oxford: Oxford University Press.

Meyer, David S., and Catherine Corrigall-Brown. 2005. "Coalitions and Political Context: U.S. Movements against Wars in Iraq." *Mobilization* 10: 327–344.

Meyer, David S., and Erin Evans. 2014. "Citizenship, Political Opportunities and Social Movements." In Handbook of Political Citizenship and Social Movements, ed. H. A. van der Heijden, 259–278. North Hampton, MA: Edward Publishing Ltd.

Meyer, David S., and Suzanne Staggenborg. 1996. "Movements, Countermovements, and the Structure of Political Opportunity." *American Journal of Sociology* 101 (6): 1628–1660.

Meyer, David S., and Sidney Tarrow, eds. 1998. *Social Movement Society: Contentious Politics for a New Century.* Lanham, MD: Rowman & Littlefield Publishers, Inc.

Meyer, John W. 2009. *World Society: The Writings of John W. Meyer.* New York: Oxford University Press.

Meyer, John W. 2010. "World Society, Institutional Theories, and the Actor." *Annual Review of Sociology*: 1–20.

McCarthy, John, and Mayer N. Zald. 1977. "Resource Mobilization and Social Movements: A Partial Theory." *American Journal of Sociology* 82 (6): 1212–1241.

Michael, Franz H. 1966. *The Taiping Rebellion.* Seattle: University of Washington Press.

Michels, Robert. 1962. *Political Parties: A Sociological Study of the Oligarchical Tendencies of Modern Democracy.* New York: Free Press.

Mies, Maria. 1986. *Patriarchy and Accumulation on A World Scale: Women in the International Division of Labour.* London: Zed Books.

Mitchell, Margaret M., Frances M. Young, and K. Scott Bowie, eds. 2006. *The Cambridge History of Christianity, Volume 1: Origins to Constantine.* Cambridge: Cambridge University Press.

Milkman, Ruth, Stephanie Luce, and Penny Lewis. 2013. "Changing the Subject: A Bottom-Up Account of Occupy Wall Street in New York City." CUNY: The Murphy Institute. http://sps.cuny.edu/filestore/1/5/7/1_a05051d2117901d/1571_92f562221b8041e.pdf.

Miller, Mary, and Karl Taube. 1993. *The Gods and Symbols of Ancient Mexico and the Maya*. London: Thames and Hudson.

Modelski, George, Tessaleno Devezas, and William R. Thompson. 2008. *Globalization as Evolutionary Process*. London: Routledge.

Moghadam, Valentine M. 2005. *Globalizing Women: Transnational Feminist Networks*. Baltimore: Johns Hopkins University Press.

Moghadam, Valentine M. 2009. "Confronting "Empire": The new imperialism, Islamism, and feminism." *Political Power and Social Theory* 20: 201–226.

Moghadam, Valentine M. 2012. "Anti-systemic Movements Compared." In *Routledge International Handbook of World-Systems Analysis*, ed. Salvatore J. Babones and Christopher Chase-Dunn. New York: Routledge.

Moghadam, Valentine M. 2013. *Globalization and Social Movements: Islamism, Feminism and the Global Justice Movement*, 2nd ed. Lanham, MD: Rowman & Littlefield.

Moghadam, Valentine M. 2017. "Explaining Divergent Outcomes of the Arab Spring: The Significance of Gender and Women's Mobilizations." *Politics, Groups, and Identities*. doi:10.1080/21565503.2016.1256824.

Moghadam, Valentine M. 2018. "Feminism and the Future of Revolutions." *Socialism and Democracy* 32 (1): 31–53. doi:10.1080/08854300.2018.1461749.

Moghadam, Valentine M. 2020a. *Globalization and Social Movements: The Populist Challenge and Democratic Alternatives*. Lanham, MD: Rowman & Littlefield.

Moghadam, Valentine M. 2020b. "Planetize the Movement! Opening Reflections for a GTI Forum." https://greattransition.org/gti-forum/planetize-movement -moghadam.

Monbiot, George. 2003. *Manifesto for a New World Order*. New York: New Press.

Moody, Kim. 1997. *Workers in a lean World: Unions in the International Economy*. London: Verso.

Mooney, James. 1965. *The Ghost Dance Religion and Sioux Outbreak of 1890*. Chicago: University of Chicago Press.

Moore, Jason W. 2015. *Capitalism in the Web of Life: Ecology and the Accumulation of Capital*. London: Verso.

Mora, Maria de Jesus, Rodolfo Rodriguez, Alejandro Zermeño, and Paul Almeida. 2018. "Immigrant Rights and Social Movements." *Sociology Compass* 12: 1–20.

Mora, Maria de Jesus, Alejandro Zermeño, Rodolfo Rodriguez, and Paul Almeida. 2017. "Exclusión y movimientos sociales en los Estados Unidos." In *Movimientos Sociales en América Latina: Perspectivas, Tendencias y Casos*, ed. P. Almeida and A. Cordero, 641–669. Buenos Aires: CLACSO.

Morgan, Hiba. 2019. "'Not Afraid of the Government': One Month of Protests in Sudan." *Al Jazeera*, January 19, https://www.aljazeera.com/news/2019 /01/afraid-government-month-protests-sudan-190119111337527.html.

Morris, Aldon. 1984. *The Origins of the Civil Rights Movement: Black Communities Organizing for Change*. New York: Free Press.

Morrow, Felix. 1974. *Revolution and Counter-Revolution in Spain*. New York: Pathfinder Press.

Mudge, Stephanie. 2018. *Leftism Reinvented: Western Parties from Socialism to Neoliberalism*. Cambridge, MA: Harvard University Press.

Mueller, Lisa. 2018. *Political Protest in Contemporary Africa*. Cambridge: Cambridge University Press.

Muis, Jasper, and Tim Immerzeel. 2017. "Causes and consequences of the rise of populist radical right parties and movements in Europe." *Current Sociology* 65 (6): 909–930.

Murphy, Craig. 1994. *International Organization and Industrial Change: Global Governance since 1850*. New York: Oxford University Press.

Nagy, Sandor. 2017. "Global Swings of the Political Spectrum since 1789: Cyclically Delayed Mirror Waves of Revolutions and Counterrevolutions." IROWS Working Paper #124, http://irows.urc.edu/papers/irows124 /irows124.htm.

Nairn, Tom. 1998. *Faces of Nationalism*. London: Verso.

Nassaney, Michael, and Kenneth E. Sassaman, eds. 1995. *Native American Interactions: Multiscalar Analyses and Interpretations in the Eastern Woodlands*. Knoxville: University of Tennessee Press.

Noël, Alain, and Jean-Philippe Thérien. 2008. *Left and Right in World Politics*. Cambridge: Cambridge University Press.

Obach, Brian K. 2004. *Labor and the Environmental Movement: The Quest for Common Ground*. Cambridge, MA: MIT Press.

O'Connor, James. 1973. *The Fiscal Crisis of the State*. New York: St. Martin's Press.

O'Connor, James. 1988. "Capitalism, Nature, Socialism: A Theoretical Introduction." *Capitalism Nature Socialism* 1 (1): 11–38.

Ortiz, David, and Sergio Béjar. 2013. "Participation in IMF-Sponsored Economic Programs and Contentious Collective Action in Latin America, 1980–2007." *Conflict Management and Peace Science* 30 (5): 492–515.

Ownby, David. 1999. "Chinese Millenarian Traditions: The Formative Age." *American Historical Review* 104 (5): 1513–1530.

Oxhorn, Philip. 2003. "Social Inequality, Civil Society, and the Limits of Citizenship in Latin America." In *What Justice? Whose Justice?: Fighting for Fairness in Latin America*, ed. S. Eckstein and T. Wickham-Crowley, 35–63. Berkeley: University of California Press.

Oxhorn, Philip. 2011. *Sustaining Civil Society: Economic Change, Democracy, and the Social Construction of Citizenship in Latin America*. State College: Pennsylvania State University Press.

Padden, R. C. 1970. *The Hummingbird and the Hawk: Conquest and Sovereignty in the Valley of Mexico, 1503–1541*. New York: Harper and Row.

Palat, Ravi. 2018. "Revolutionary Transformations in Asia, 1840–1873: Changing Relational Flows, Production Processes, and Social Upheavals." Presented at the annual meeting of the Social Science History Association, Phoenix, AZ.

Paret, Marcle. 2017. Southern Resistance in Critical Perspective. In *Southern Resistance in Critical Perspective*, ed. M. Paret, C. Runciman, and L. Sinwell, 1–18. New York: Routledge.

Passmore, K. 2012. *Fascism: A Very Short Introduction*. Oxford: Oxford University Press.

Patel, Raj, and Philip McMichael. 2009. "A Political Economy of the Food Riot." *Review* 32 (1): 9–35.

Patomäki, Heikki. 2008. *The Political Economy of Global Security*. New York: Routledge.

Patomäki, Heikki. 2019. "A World Political Party: The Time Has Come." *Great Transition Network Essay*, February, https://www.greattransition.org/.

Pauketat, Timothy. 2009. *Cahokia: Ancient America's Great City on the Mississippi*. New York: Viking.

Paulsen, Ronnelle, and Karen Glumm. 1995. "Resource Mobilization and the Importance of Bridging Beneficiary and Conscience Constituencies." *National Journal of Sociology* 9 (2): 37–62.

Paxton, Robert O. 2004. *The Anatomy of Fascism*. New York: Vintage Books.

Pellow, David Naguib. 2017. *What Is Critical Environmental Justice?* London: Polity Press.

Pereyra, Sebastián, Germán Perez, and Federico Schuster. 2015. "Trends of Social Protest in Argentina: 1989–2007." In *Handbook of Social Movements across Latin America*, ed. P. Almeida and A. Cordero, 335–360. New York: Springer.

Picketty, Thomas. 2014. *Capital in the 21st Century*. Cambridge, MA: Harvard University Press.

Piven, Frances Fox, and Richard Cloward. 1977. *Poor People's Movements: Why They Succeed, How They Fail*. New York: Vintage.

Pleyers, Geoffrey. 2010. *Alter-Globalization*. Cambridge, MA: Polity Press.

Polanyi, Karl. 1944. *The Great Transformation*. Boston: Beacon Press.

Polasky, Janet. 2016. *Revolutions Without Borders: The Call to Liberty in the Atlantic World*. New Haven, CT: Yale University Press.

Postgate, J. N. 1992. *Early Mesopotamia: Society and Economy at the Dawn of History*. New York: Routledge.

Poulantzas, Nicos. 1974. *Fascism and Dictatorship: The Third International and the Problem of Fascism*. London: New Left Books.

Prashad, Vijay. 2007. *The Darker Nations: A People's History of the Third World*. New York: The New Press.

Prehofer, Christian, and Christian Bettstetter. 2005. "Self-Organization in Communication Networks: Principles and Design Paradigms." *IEEE Communications Magazine* 43 (7): 78–85

Ramirez, F., Y. Soysal, and S. Shanahan. 1997. "The Changing Logic of Political Citizenship: Cross-National Acquisition of Women's Suffrage Rights, 1890 to 1990." *American Sociological Review* 62 (5): 735–745. http://www.jstor.org/stable/2657357.

Ramiro, Luis, and Raul Gomez. 2017. "Radical-Left Populism during the Great Recession: *Podemos* and Its Competition with the Established Radical Left." *Political Studies* 65 (1S): 108–126.

Raventós, Ciska. 2018. *Mi Corazón Dice No: El Movimiento de Oposición al TLC*. San José, Costa Rica: Editorial Universidad de Costa Rica.

Reitan, Ruth. 2007. *Global Activism*. London: Routledge.

Reitan, Ruth. 2012a. "Introduction: Theorizing and Engaging the Global Movement: From Anti-Globalization to Global Democratization." *Globalizations* 9 (3): 323–335.

Reitan, Ruth. 2012b. "Coalescence of the Global Peace and Justice Movements." *Globalizations* 9 (3): 337–350.

Reitan, Ruth, and Shannon Gibson. 2012. "Climate Change or Social Change? Environmental and Leftist Praxis and Participatory Action Research." *Globalizations* 9 (3): 395–410.

Richardson, Seth F. C., ed. 2010. "Rebellions and Peripheries in the Cuneiform World." *American Oriental Series* 91. Ann Arbor, MI: American Oriental Society.

Riley, Dylan. 2018. "What Is Trump?" *New Left Review* 114: 5–31.

Roberts, Bryan R., and Alejandro Portes. 2006. "Coping with the Free Market City: Collective Action in Six Latin American Cities at the End of the Twentieth Century." *Latin American Research Review* 41 (2): 57–83.

Roberts, J. T., and Bradley Parks. 2006. *A Climate of Injustice: Global Inequality, North-South Politics, and Climate Policy*. Cambridge, MA: MIT Press.

Roberts, J. T., D. Pellow, and P. Mohai. 2018. "Environmental Justice." In *Environment and Society*, 233–255. New York: Palgrave Macmillan.

Roberts, Kenneth. 2008. "The Mobilization of Opposition to Economic Liberalization." *Annual Review of Political Science* 11: 327–349.

Robinson, William I. 1996. *Promoting Polyarchy: Globalization, US Intervention and Hegemony*. Cambridge: Cambridge University Press.

Robinson, William I. 2004. *A Theory of Global Capitalism: Production, Class, and State in a Transnational World*. Baltimore: Johns Hopkins University Press.

Robinson, William I. 2008. *Latin America and Global Capitalism*. Baltimore: Johns Hopkins University Press.

Robinson, William I. 2013. "Policing the Global Crisis." *Journal of World-Systems Research* 19 (2): 193–197.

Robinson, William I. 2014. *Global Capitalism and the Crisis of Humanity*. Cambridge: Cambridge University Press.

Robinson, William I. 2019. *Into the Tempest: Essays on the New Global Capitalism*. Chicago: Haymarket.

Robinson, William I., and M. Barrera. 2012. "Global Capitalism and Twenty-First Century Fascism: A U.S. Case Study." *Race & Class* 53 (3), 4–29.

Rodney, Walter. 1981. *How Europe Underdeveloped Africa*. Washington, DC: Howard University Press.

Rodrik, Dani. 2018. "Populism and the Economics of Globalization." *Journal of International Business Policy*. doi:10.1057/s42214-018-001-4.

Rojas, Fabio. 2017. *Theory for the Working Sociologist*. New York: Columbia University Press.

Romani, John. F. 1981. "Astronomy and Social Integration: An Examination of Astronomy in a Hunter and Gatherer Polity." Thesis submitted in partial satisfaction of the requirements for the degree of MA in Anthropology, California State University, Northridge. http://scholarworks.csun.edu/handle /10211.3/127672.

Rose, Fred. 2000. *Coalitions Across the Class Divide: Lessons from the Labor, Peace, and Environmental Movements*. Ithaca, NY: Cornell University Press.

Ross, Robert J. S., and Kent Trachte. 1990. *Global Capitalism: The New Leviathan*. Albany: State University of New York Press.

Rossi, Federico. 2017. *The Poor's Struggle for Political Incorporation: The Piquetero Movement in Argentina*. Cambridge: Cambridge University Press.

Rostow, W. W. 1960. *The Stages of Economic Growth*, 3rd ed. Cambridge: Cambridge University Press.

Rubin, Beth. 2012. "Shifting Social Contracts and the Sociological Imagination." *Social Forces* 91 (2): 327–346.

Ruby, Robert H., and John A. Brown. 1989. *Dreamer-Prophets of the Columbia Plateau: Smohalla and Skolaskin*. Norman, OK: University of Oklahoma Press.

Sahlins, Marshall. 1972. *Stone Age Economics*. Chicago: Aldine.

Sanbonmatsu, John. 2004. *The Postmodern Prince: Critical Theory, Left Strategy and the Making of a New Political Subject*. New York: Monthly Review Press.

Sanderson, Stephen K. 1995. *Social Transformations*. London: Basil Blackwell.

Sanderson, Stephen K. 2018. *Religious Evolution and the Axial Age*. London: Bloomsbury.

Sandoval, Salvador. 2007. "Alternative forms of working-class organization and the mobilization of informal-sector workers in Brazil in the era of neoliberalism." *International Labor and Working-Class History* 72 (1): 63–89.

Santiago-Valles, Kelvin. 2005. "World Historical Ties Among "Spontaneous" Slave Rebellions in the Atlantic." *Review* 28 (1): 51–84.

Santos, Boaventura de Sousa. 2006. *The Rise of the Global Left*. London: Zed Books.

Sassaman, J. A. 2005. "Poverty Point as Structure, Event, Process." *Journal of Archaeological Method and Theory* 12 (4): 335–364. http://link.springer.com/article/10.1007/s10816-005-8460-4.

Sassen, Saskia. 2008. *Territory, Authority, Rights: From Medieval to Global Assemblages*, 2nd ed. Princeton: Princeton University Press.

Saull, Richard G. 2015. "Capitalist Development and the Rise and 'Fall' of the Far-Right." *Critical Sociology* 41 (4–5): 619–639.

Saull, Richard G., Alexander Anievas, Neil Davidson, and Adam Fabry, eds. 2015. *The Longue Durée of the Far Right*. London: Routledge.

Saunders, Joe W., Rolfe D. Mandel, C. Garth Sampson, Charles M. Allen, E. Thurman Allen, Daniel A. Bush, James K. Feathers, Kristen J. Gremillion, C. T. Hallmark, H. Edwin Jackson, Jay K. Johnson, Reca Jones, Roger T. Saucier, Gary L. Stringer, and Malcolm F. Vidrine. 2005. "Watson Brake, a Middle Archaic Mound Complex in Northeast Louisiana." *American Antiquity* 70 (4): 631–668. http://www.jstor.org/stable/40035868?seq=1#page_scan_tab_contents.

Scarborough, Vernon L., and David R. Wilcox, eds. 1991. *The Mesoamerican Ballgame*. Tucson: University of Arizona Press.

Schaeffer, Robert K. 2014. *Social Movements and Global Social Change*. Lanham, MD: Rowman & Littlefield.

Schindler, Seth, and Juan Miguel Kanai. 2019. "How mega infrastructure projects in Africa, Asia and Latin America are reshaping development." *The Conversation*, October 31. https://theconversation.com/how-mega-infrastructure-projects-in-africa-asia-and-latin-america-are-reshaping-development-125449.

Schnaiberg, Allan. 1980. *The Environment: From Surplus to Scarcity*. Oxford: Oxford University Press.

Schock, Kurt. 2005. *Unarmed Insurrections: People Power Movements in Nondemocracies*. Minneapolis: University of Minnesota Press.

Schofer, Evan, and John W. Meyer. 2005. "The Worldwide Expansion of Higher Education in the Twentieth Century." *American Sociological Review* 70 (6): 898–920.

Scott, James C. 1998. *Seeing like a State: How Certain Schemes to Improve the Human Condition Have Failed.* New Haven, CT: Yale University Press.

Scott, James C. 2017. *Against the Grain: A Deep History of the Earliest States.* New Haven, CT: Yale University Press.

Segura-Ubiergo, Alex. 2007. *The Political Economy of the Welfare State in Latin America: Globalization, Democracy, and Development.* Cambridge: Cambridge University Press.

Shefner, Jon and Cory Blad. 2020. *Why Austerity Persists.* Cambridge, UK: Polity Press.

Silva, Eduardo. 2009. *Challenges to Neoliberalism in Latin America.* Cambridge: Cambridge University Press.

Silver, Beverly. 2003. *Forces of Labor: Workers' Movements and Globalization since 1870.* Cambridge: Cambridge University Press.

Simmons, Erica. 2014. "Grievances do Matter in Mobilization." *Theory and Society* 43: 513–546.

Simmons, Erica. 2016. *Meaningful Mobilization: Market Reforms and the Roots of Social Protest in Latin America.* Cambridge: Cambridge University Press.

Skinner, G. William. 1971. "Chinese Peasants and the Closed Community: An Open and Shut Case." *Comparative Studies in Society and History* 13: 270–281.

Sklair, Leslie. 2001. *The Transnational Capitalist Class.* Malden, MA: Blackwell.

Skocpol, Theda, and Alexander Hertel-Fernandez. 2016. "The Koch network and republican party extremism." *Perspectives on Politics* 14 (3): 681–699.

Skocpol, Theda, and Vanessa Williamson. 2016. *The Tea Party and the Remaking of Republican Conservatism.* New York: Oxford University Press.

Smelser, Neil J. 1962. *Theory of Collective Behavior.* New York: The Free Press.

Smith, Jackie. 2001. "Globalizing Resistance: The Battle of Seattle and the Future of Social Movements." *Mobilization* 6 (1): 1–21.

Smith, Jackie. 2008. *Social Movements for Global Democracy.* Baltimore: Johns Hopkins University Press.

Smith, Jackie. 2014. "Counter-Hegemonic Networks and the Transformation of Global Climate Politics: Rethinking Movement-State Relations." *Global Discourse* 4 (2–3): 120–138.

Smith, Jackie, Basak Gemici, Melanie M. Hughes, and Samantha Plummer. 2018. "Transnational Social Movement Organizations and Counterhegemonic Struggle Today." *Journal of World Systems Research* 24 (2): 372–403.

Smith, Jackie, Marina Karides, Marc Becker, Dorval Brunelle, Christopher Chase-Dunn, Donatella della Porta, Rosalba Icaza Garza, Jeffrey S. Juris, Lorenzo

Mosca, Ellen Reese, Peter Jay Smith, and Rolando Vazquez. 2014. *Global Democracy and the World Social Forums*, revised 2nd ed. Boulder, CO: Paradigm Publishers.

Smith, Jackie, and Dawn Wiest. 2012. *Social Movements in the World-System: The Politics of Crisis and Transformation*. New York: Russell Sage Foundation.

Smoak, Gregory E. 2006. *Ghost Dances and Identity: Prophetic Religion and American Indian Ethnogenesis in the Nineteenth Century*. Berkeley: University of California Press.

Sneyd, Lauren Q., Alexander Legwegoh, and Evan D. G. Fraser. 2013. "Food riots: Media perspectives on the causes of food protest in Africa." *Food Security* 5 (4): 485–497.

Snow, David A., and Colin Bernatzky. 2019. "The Coterminous Rise of Right-Wing Populism and Superfluous Populations." In *Populism and the Crisis of Democracy, Volume 1: Concepts and Theory*, ed. Grego Fitzi, Jurgen Mackert and Bryan S. Turner. London: Routledge.

Snow, David A., Daniel Cress, Liam Downey, and Andrew Jones. 1998. "Disrupting the 'Quotidian': Reconceptualizing the Relationship Between Breakdown and the Emergence of Collective Action." *Mobilization* 3 (1): 1–22.

Snow, David A., and Peter B. Owens. 2014. "Social Movements and Social Inequality: Toward a More Balanced Assessment of the Relationship." In *Handbook of the Social Psychology of Inequality*, ed. J. D. McLeod, E. J. Lawler, and M. Schwalbe, 657–681. New York: Springer.

Snow, David A., and Sarah A. Soule. 2010. *A Primer on Social Movements*. New York: Norton.

Snow, David A., S. A. Soule, H. Kriesi, and H. J. McCammon, eds. 2018. "Introduction: Mapping and Opening Up the Terrain." In *The Wiley Blackwell Companion to Social Movements*, 1–16. doi:10.1002/9781119168577.

Somers, Margaret. 2008. *The Right to Have Rights*. Cambridge: Cambridge University Press.

Somma, Nicolas, Matías Bargsted, Rodolfo Disi Pavlic, and Rodrigo M. Medel. 2020. "No Water in the Oasis: The Chilean Spring of 2019-2020." *Social Movement Studies* (published online February 11). doi:10.1080/14742837.2020.1727737.

Sosa, Eugenio. 2013. *Dinámica de la protesta social en Honduras*. Tegucigalpa: Editorial Guaymuras.

Sosa, Eugenio, and Paul Almeida. 2019. "Honduras: A Decade of Popular Resistance." *NACLA Report on the Americas* 51 (4).

Spalding, Rose. 2014. *Contesting Trade in Central America: Market Reform and Resistance*. Austin: University of Texas Press.

Spektorowski, Alberto. 2016. "Fascism and Post-National Europe." *Theory, Culture & Society* 33 (1): 115–138.

Spence, Jonathan D. 1996. *God's Chinese Son: the Taiping Heavenly Kingdom of Hong Xiuquan*. New York: Norton.

Spier, Leslie. 1921. "The Sun Dance of the Plains Indians: Its Development and Diffusion." *New York: Anthropological Papers of the American Museum of Natural History* 16 (Part 2).

Spier, Leslie. 1927. *The Ghost Dance of 1870 among the Klamath of Oregon*. Seattle: University of Washington Press.

Spier, Leslie. 1935. "The Prophet Dance of the Northwest and its Derivatives: The Source of the Ghost Dance." *General Series in Anthropology* (1). Menasha, WI: George Banta Publishing Company.

Spronk, Susan, and Philipp Terhorst. 2012. "Social Movement Struggles for Public Services." In *Alternatives to Privatization in the Global South*, ed. David McDonald and Greg Ruiters, 133–156. New York: Routledge.

Standing, Guy. 2011. *The Precariat: The New Dangerous Class*. New York: Bloomsbury Academic.

Standing, Guy. 2014. *A Precariat Charter: From Denizens to Citizens New York*. New York: Bloomsbury.

Stark, Rodney. 1996. *The Rise of Christianity: How the Obscure, Marginal Jesus Movement Became the Dominant Religious Force in the Western World in a Few Centuries*. Princeton, NJ: Princeton University Press.

Starr, Amory. 2000. *Naming the Enemy: Anti-corporate Movements Confront Globalization*. London: Zed Books.

Steger, Manfred, James Goodman, and Erin K. Wilson. 2013. *Justice Globalism: Ideology, Crises, Policy*. Thousand Oaks, CA: Sage.

Strang, D., and S. A. Soule. 1998. "Diffusion in Organizations and Social Movements: From Hybrid Corn to Poison Pills." *Annual Review of Sociology* 24: 265–290.

Struna, Jason. 2013. "Global Capitalism and Transnational Class Formation." *Globalizations* 10 (5): 651–658.

Subramaniam, Mangala. 2007. "NGOs and resources in the construction of intellectual realms: cases from India." *Critical Sociology* 33 (3): 551–573.

Suttles, Wayne. 1987. *Coast Salish Essays*. Seattle: University of Washington Press.

Svampa, Maristella, and Sebastián Pereyra. 2009. *Entre la ruta y el barrio: la experiencia de las organizaciones piqueteras*, 3rd ed. Buenos Aires: Editorial Biblos.

Swanson, Guy E. 1960. *The Birth of the Gods: The Origin of Primitive Beliefs*. Ann Arbor: University of Michigan Press.

Sworakowski, Witold S. 1965. *The Communist International and Its Front Organizations*. Stanford, CA: Hoover Institution on War, Revolution, and Peace.

Tarrow, Sidney. 2005. *The New Transnational Activism*. Cambridge: Cambridge University Press.

Tarrow, Sidney. 2011. *Power in Movement: Social Movements, Collective Action, and Politics*, 3rd ed. Cambridge: Cambridge University Press.

Tarrow, Sidney. 2018. "Rhythms of Resistance: The Anti-Trumpian Moment in a Cycle of Contention." In *The Resistance: The Dawn of the Anti-Trump Opposition Movement*, ed. David Meyer and Sidney Tarrow, 187–206. Oxford: Oxford University Press.

Tarrow, Sidney, and Doug McAdam. 2005. "Scale Shift in Transnational Contention." In *Transnational Protest and Global Activism*, ed. Donatella della Porta and Sidney Tarrow, 121–150. Boulder, CO: Rowman & Littlefield.

Tattersall, Amanda. 2010. *Power in Coalition*. Ithaca, NY: Cornell University Press.

Taylor-Gooby, Peter. 2009. *Reframing Social Citizenship*. Oxford: Oxford University Press.

Terriquez, Veronica, Tizoc Brenes, and Abdiel Lopez. 2018. "Intersectionality as a Multipurpose Collective Action Frame: The Case of the Undocumented Youth Movement." *Ethnicities* 18 (2): 260–276.

Thornton, Russell. 1981. "Demographic Antecedents of a Revitalization Movement: Population Change, Population Size and the 1890 Ghost Dance." *American Sociological Review* 46 (1): 88–96.

Thornton, Russell. 1986. *We Shall Live Again: The 1870 and 1890 Ghost Dance Movements as Demographic Revitalization*. Cambridge: Cambridge University Press.

Thunberg, Greta. 2019. *No One Is Too Small to Make a Difference*. New York: Penguin Books.

Tilly, Charles. 1964. *The Vendee*. Cambridge, MA: Harvard University Press.

Tilly, Charles. 1978. *From Mobilization to Revolution*. Reading, MA: Addison-Wesley.

Tilly, Charles. 1984. "Social Movements and National Politics." In *State Making and Social Movements*, ed. C. Bright and S. Harding, 297–317. Ann Arbor: University of Michigan Press.

Tilly, Charles. 1989. *Big Structures, Large Processes, Huge Comparisons*. New York: Russell Sage.

Tilly, Charles. 2003. *The Politics of Collective Violence*. Cambridge: Cambridge University Press.

Tilly, Charles, and Sidney Tarrow. 2015. *Contentious Politics*. New York: Oxford University Press.

Tilly, Charles, and Lesley Wood. 2013. *Social Movements, 1768–2004*. Boulder, CO: Paradigm.

Tong, Y., and S. Lei. 2013. *Social Protest in Contemporary China, 2003–2010: Transitional Pains and Regime Legitimacy*. London: Routledge.

Traverso, Enzo. 2017. *The New Faces of Fascism: Populism and the Far Right*. London: Verso.

Trejo, Guillermo. 2012. *Popular Movements in Autocracies: Religion, Repression, and Indigenous Collective Action*. New York: Cambridge University Press.

Turchin, Peter. 2003. *Historical Dynamics*. Princeton, NJ: Princeton University Press.

Turchin, Peter. 2016a. *Ultrasociety*. Chaplin, CT: Beresta Books.

Turchin, Peter. 2016b. *Ages of Discord: A Structural-Demographic Analysis of American History*. Chaplin, CT: Beresta Books.

Turchin, Peter, and Sergey A. Nefedov. 2009. *Secular Cycles*. Princeton, NJ: Princeton University Press.

Uba, Katrin. 2005. "Political Protest and Policy Change: The Direct Impacts of Indian Anti-Privatization Mobilizations, 1990–2003." *Mobilization: An International Quarterly* 10 (3): 383–396.

Uba, Katrin. 2008. "Labor Union Resistance to Economic Liberalization in India: What Can National and State Level Patterns of Protests against Privatization Tell Us?" *Asia Survey* 48 (5): 860–884.

Van de Mieroop, Marc. 1999. *The Ancient Mesopotamian City*. New York: Oxford University Press.

Van de Mieroop, Marc. 2011. *A History of Ancient Egypt*. Malden, MS: Wiley-Blackwell.

Van der Veer, P. 2002. "Colonial Cosmopolitanism." In *Conceiving Cosmopolitanism*, ed. S. Vertovec and R. Cohen, 165–179. New York: Oxford University Press.

Van Dyke, Nella. 2003. "Crossing Movement Boundaries: Factors That Facilitate Coalition Protest by American College Students, 1930–1990." *Social Problems* 50 (2): 226–550. http://www.jstor.org/stable/10.1525/sp.2003.50.2.226.

Van Dyke, Nella, and Bryan Amos. 2017. "Social Movement Coalitions: Formation, Longevity, and Success." *Sociology Compass* 11 (7): 1–17.

Van Dyke, Nella, and Holly J. McCammon, eds. 2010. *Strategic Alliances: Coalition Building and Social Movements*. Minneapolis: University of Minnesota Press.

Van Dyke, Nella, and Sarah A. Soule. 2002. "Structural Social Change and the Mobilizing Effect of Threat: Explaining Levels of Patriot and Militia Organizing in the United States." *Social Problems* 49 (4): 497–520.

Vasi, Ion Bogdan, and Chan S. Suh. 2016. "Online Activities, Spatial Proximity, and the Diffusion of the Occupy Wall Street Movement in the United States." *Mobilization* 21 (2): 139–154.

von Bülow, Marisa. 2011. *Building Transnational Networks: Civil Society and the Politics of Trade in the Americas.* Cambridge: Cambridge University Press.

Vreeland, James. 2003. *The IMF and Economic Development.* Cambridge: Cambridge University Press.

Vreeland, James. 2007. *The International Monetary Fund: Politics of Conditional Lending.* New York: Routledge.

Wagar, W. Warren. 1992. *A Short History of the Future.* Chicago: University of Chicago Press.

Wallace, Anthony F. C. 1956. "Revitalization Movements." *American Anthropologist* 58: 264–281.

Wallace, Anthony F. C. 1965. "James Mooney (1861–1921) and the Study of the Ghost Dance Religion." In *The Ghost-Dance Religion and the Sioux Outbreak of 1890*, ed. James Mooney. Chicago: University of Chicago Press.

Wallerstein, Immanuel. 1974. *The Modern World-System*, vol. 1. New York: Academic Press.

Wallerstein, Immanuel. 1984. "The Three Instances of Hegemony in the History of the Capitalist World-Economy." In *Current Issues and Research in Macrosociology, International Studies in Sociology and Social Anthropology*, vol. 37, ed. Gerhard Lenski, 100–108. Leiden, Netherlands: E. J. Brill.

Wallerstein, Immanuel. 1990. "Antisystemic Movements: History and Dilemmas." In *Transforming the Revolution*, ed. Samir Amin, Giovanni Arrighi, Andre Gunder Frank, and Immanuel Wallerstein. New York: Monthly Review Press.

Wallerstein, Immanuel. 2003a. "Entering Global Anarchy." *New Left Review* 22: 27–35.

Wallerstein, Immanuel. 2003b. *The Decline of American Power.* New York: New Press.

Wallerstein, Immanuel. 2004. *World-Systems Analysis.* Durham, NC: Duke University Press.

Wallerstein, Immanuel. 2008. "The Political Construction of Islam." In *Islam and the Orientalist World-System*, ed. K. Samman and M. Al-Zo'by. Boulder, CO: Paradigm.

Wallerstein, Immanuel. 2011. *The Modern World-System Volume 4: Centrist Liberalism Triumphant 1789–1914.* Berkeley: University of California Press.

Wallerstein, Immanuel. 2012. *The Modern World-System*, vol. 4. Berkeley, CA: University of California Press.

Wallerstein, Immanuel, ed. 2017. "The Political Construction of Islam." In *The World-System and Africa*, 125–140. New York: Diasporic Africa Press.

Wallerstein, Immanuel, Randall Collins, Michael Mann, Georgi Derlugian, and Craig Calhoun. 2013. *Does Capitalism Have a Future?* New York: Oxford University Press.

Walton, John, and Charles Ragin. 1990. "Global and National Sources of Political Protest: Third World Responses to the Debt Crisis." *American Sociological Review* 55 (6): 876–891.

Walton, John, and David Seddon. 1994. *Free Markets and Food Riots.* Oxford: Blackwell.

Walton, John, and Jonathan Shefner. 1994. "Latin America: Popular Protest and the State." In *Free Markets and Food Riots: The Politics of Global Adjustment*, ed. J. Walton and D. Seddon, 97–134. Oxford: Blackwell.

Waterman, Peter. 2006. "Toward a global labour charter for the 21st century." https://laborstrategies.blogs.com/global_labor_strategies/global_unionism/page/4/.

Wheatley, Paul. 1975. "Satyanrta Suvarnadvipa: From Reciprocity to Redistribution in Ancient Southeast Asia." In *Ancient Civilization and Trade*, ed. J. A. Sabloff and C. C. Lamberg-Karlovski, 227–284. Albuquerque: University of New Mexico Press.

White, James W. 1995. *Ikki: Social Conflict and Political Protest in Early Modern Japan.* Ithaca, NY: Cornell University Press.

Whiteneck, Daniel J. 1996. "The Industrial Revolution and Birth of the Anti-Mercantilist Idea:Epistemic Communities and Global Leadership." *Journal of World-Systems Research* 2 (1): 2–35. doi: 10.5195/jwsr.1996.69.

Wolf, Eric R. 1997. *Europe and the People Without History.* Berkeley: University of California Press.

Wood, Lesley J. 2004. "Breaking the Bank and Taking to the Streets: How Protesters Target Neoliberalism." *Journal of World-Systems Research* 10 (1): 69–89.

Wood, Lesley J. 2012. *Direct Action, Deliberation, and Diffusion: Collective Action After the WTO Protests in Seattle.* Cambridge: Cambridge University Press.

Woodhouse, Keith M. 2018. *The Ecocentrists.* New York: Columbia University Press.

World Meteorological Organization. 2018. *WMO Provisional statement on the State of the Global Climate in 2018.* New York: United Nations.

World Social Forum. 2001. *Charter of Principles.* http://www.universidadepopular.org/site/media/documentos/WSF_-_charter_of_Principles.pdf.

Worsley, Peter. 1968. *The Trumpet Shall Sound: A Study of "Cargo" Cults in Melanesia.* New York: Schocken.

Wright, Erik O. 2010. *Envisioning Real Utopias.* London: Verso.

Yanacopulos, Helen. 2007. "Cutting the Diamond: Networking Economic Justice." Paper presented at the Institute for International, Comparative, and Area Studies (IICAS), University of California, San Diego, January 25–27.

Yanacopulos, Helen, and Matt Baillie Smith. 2008. "The Ambivalent Cosmopolitanism of International NGOs." In *Can NGOs Make a Difference? The Challenge of Development Alternatives*, ed. Anthony Bebbington, Samual Hickey, and Diana Mitlin, 298–315. London: Zed Books.

Youngs, Richard. 2017. "What Are the Meanings Behind the Worldwide Rise in Protest?" *openDemocracy* (October). https://www.opendemocracy.net/pro test/multiple-meanings-global-protest.

Zagarell, Allen. 1986. "Trade, Women, Class and Society in Ancient Western Asia." *Current Anthropology* 27 (5): 415–430.

Zald, Mayer N., and John D. McCarthy. 1987. "Social Movement Industries: Competition and Conflict." In *Social Movements in and Organizational Society*, ed. Mayer N. Zald and John D. McCarthy, 161–184. New Brunswick, NJ: Transactions Publishers.

Zald, Mayer N., and Bert Useem. 1987. "Movement and Countermovement Interaction: Mobilization, Tactics, and State Involvement." In *Social Movements in an Organizational Society*, ed. Mayer N. Zald and John D. McCarthy, 247–272. New York: Routledge.

Zhang, David D., C. Y. Jim, George C-S. Lin, Yuan-Qing He, James J. Wang, and Harry F. Lee. 2006. "Climatic Change, Wars and Dynastic Cycles in China over the Last Millennium." *Climatic Change* 76: 459–477. https://link .springer.com/article/10.1007/s10584-005-9024-z.

Zhang, David D., Qing Pei, Harry F. Lee, Jane Zhang, Chun Qi Chang, Baosheng Li, Jinbao Li, and Xiaoyang Zhang. 2015. "The Pulse of Imperial China: A Quantitative Analysis of Long-Term Geopolitical and Climatic Cycles." *Global Ecology and Biogeography* 24 (1): 87–96. doi:10.1111/geb.12247.

Zhang, David D., Jane Zhang, Harry F. Lee, and Yuan-Qing He. 2007. "Climate Change and War Frequency in Eastern China Over the Last Millennium." *Human Ecology* 35: 403–414. https://link.springer.com/article/10 .1007%2Fs10745-007-9115-8.

Zibechi, Raúl. 2010. *Dispersing Power*. Oakland, CA: AK Books.

Zuboff, Shoshana. 2019. *The Rise of Surveillance Capitalism*. New York: Hachette.

Index

58–59, 64–65, 66, 67–68, 69;
anti-privatization activities, 63; fascist
attacks on, 129–30; as social
movement, 61, 112–13; state sectors
in, 48, 49; teachers', 54, 67–68
Latin America: anti-privatization
movements, 62–63, 69; debt crisis, 49,
50, 130; democracy, 135–36;
independence struggles, 89; oppositional political parties, 60; Pink Tide,
90, 92, 107, 130, 131–32, 133,
134–36; transnational protest
campaigns, 69
leaders: charismatic, 27, 33, 120, 136;
social classes of, 20
League of Nations, 122, 163n8
"lean state" ideology, 5
Lebanon, 58, 147
left-wing populism, 131
Lenin, Vladimir, 105, 116, 129
LGBTQ rights, 92, 104
Liberation Theology, 146
Lineages of the Absolutist State
(Anderson), 124
Living wage campaigns, 61
local level, of social movements, 62–68,
71, 93, 147–48
love, 115, 116
Lula da Silva, Luiz, 108

Mann, Michael, 125–26, 127
Maoists, 106
marcher states, 30–31
Marx, Karl, 111
Marxism, 94, 97, 127, 133, 156n2
mass mobilizations, 49–61, 65–66, 76,
79, 134, 147; in 2011–2012, 90,
133–34; Climate Justice movement,
80–81, 82–85; examples, 51–61;
fascist/neofascist, 127, 129–30, 141;
largest, 74, 78, 148; right-wing,
139–40; social citizenship threats and,
49–51; spontaneity, 136–37; state-led
development sector in, 51–57, 58–59,
61, 63–68, 147–48
materialism, 140
media, alternative, 93, 121. *See also*
social media
meritocracy, 119

Mesoamerica, religious ideologies,
22–28
Mesopotamia, 28–29, 30, 33–34, 142
metahumanism, 115–17
Me-Too movement, 133–34
Mexican Revolution, 89
Mexico, 56, 60, 62–63, 89, 147;
Zapatista rebellion, 56, 62, 69, 90, 99,
108
Middle East and North Africa (MENA),
57–58, 90, 134
migrants/immigrants, 130, 131, 132,
134
migration, 6, 16, 18, 21, 28, 104, 112,
131, 132, 134, 140–41, 146. *See also*
emigration
military coups, 120, 135
military expansionism, 138–39
millenarian movements, 24–26, 109,
157nn14–15
Morales, Evo, 145–46
moral order, 16, 19, 115–16, 142
morphological discourse analysis, 98–99
"Mother Earth" (Pachamama), 84–85
movement articulation, 99, 103, 112–17
Movement Toward Socialism (MAS),
145–46
multiculturalism, 139, 140
Mussolini, Benito, 126–27

nationalism, 120, 151
national liberation movements, 92, 129
nation-states, 44, 129; theocratic, 27–30
neofascism, 14, 106, 125–26, 134, 136,
137, 138–39
neoliberalism, 4, 9–10, 39, 49, 51, 121;
policies, 68–71
neoliberalism resistance. *See* anti-
neoliberalism movements
Netherlands, 88, 89
New Deal, 88, 109
New Global Left, 87–117; coalitions,
92–98, 103–7; Global Justice
movement, 98–99
New Left, 5, 89, 106, 108, 133, 137
Nicaragua, 55, 56, 60, 63, 135
Nigeria, 57, 63, 69
nongovernmental organizations
(NGOs), 62–63, 74, 80–81, 97, 98

normative regulation, 22
norms, consensual, 15, 26, 123
North American Free Trade Agreement (NAFTA), 56, 69

Obrador, López, 136
Occupy movements, 69, 113, 114, 133; Occupy Wall Street, 64, 78, 90, 99, 109
Old Left, 89, 90, 106, 108, 113, 115, 137
ontology, 118
open source knowledge, 93
oppression, 115, 116, 153

Pachamama ("Mother Earth"), 84–85
Panama, 55, 63
Paris Climate Accords, 1
Paris Climate Agreement, 81, 82
peace movements, 92, 102, 104, 121
peripheralization, 110
Peru, 69, 147
philosophy, postmodern, 94
political parties, 2, 7, 42, 44, 56, 122, 133; Catholic, 165n22; global, 8; leftist oppositional, 57, 59, 60–61, 62–63, 64–65, 69, 130; progressive, 133
polyarchy, 119
popular fronts. See united and popular fronts
population density, 16, 19, 21, 26, 140, 143
populism, 130. See also left-wing populism; right-wing populism
Port Huron statement, 89, 90
post-capitalism, 114–15
"post-fascism," 140
postmodern prince, 115–16
power, coercive, 15–16
precariat, 110–13
prefiguration, 99, 107, 116–17
premodern collective action/social movements, 11–12, 15–38, 145–46; evidence, 15, 21, 22; in small-scale polities, 21–30
premodern societies: sociocultural evolution, 21–22; tributary accumulation, 123–24
price controls/mobilizations, 46–47, 52–53, 54–55, 56, 57–58, 70, 147

privatization, 5, 18, 53, 55, 57, 58–59, 62–63, 69
proactive social movements, 17–18, 35–36
Progressive International, 116–17
Prophet Dance, 25–26, 27
Protestant Reformation, 8, 38, 89, 132
protest campaigns, social sectors in, 48–49. See also mass mobilizations

Qing dynasty, 18, 35

racism, 86, 89, 100–102, 104, 114, 116–17, 131, 132, 140, 146, 151
reactive social movements, 17–18, 36
Reagan, Ronald, 5
rebellions, 16, 87, 88–89; Taiping, 34, 89; Zapatista, 56, 62, 69, 90, 99, 108
recessions, 43, 61, 64
relative deprivation theory, 40
religion/spirituality: as agent of social change, 17, 146; fascism and, 126; hierarchical, 142; mass heterodox movements, 34–35; in premodern states, 12, 22, 27–30; in right-wing social movements, 132; rise of, 32–36
religious fundamentalism, 107, 122, 132–33, 134
repertoires of contention, 95–96
repressive regimes, 142
resource curse, 135
revitalization movements, 34; Native American/Ghost Dance, 12, 22–25, 26, 27
revolutions, in ancient world, 30–38. See also World Revolution
Revolutions: A Very Short Introduction (Goldstone), 30
right-wing movements, socially constructed threats of, 10
right-wing populism, 11, 14, 118, 125, 126–27, 130–31, 135, 136, 137, 140, 141, 142–43, 151, 154
rituals, 16, 27–30; Ghost Dance, 12, 22–25, 26, 27
Rome, ancient, 31, 33–34, 124
Russian Revolution, 89